Food System Sustainability

Insights from duALIne

As Western-style food systems extend further around the world, food sustainability is becoming an increasingly important issue. Such systems are not sustainable in terms of their consumption of resources, their impact on ecosystems or their effect on health and social equality.

From 2009 to 2011, the duALIne project, led by INRA and CIRAD, assembled a team of experts to investigate food systems downstream of the farm, from the farm gate to consumption. Representing a diverse range of backgrounds, spanning academia and the public and private sectors, the project aimed to review the international literature and to take account of the experts' own expertise to identify major gaps in our knowledge. This book brings together its key conclusions and insights, presents state-of-the-art research in food sustainability and identifies priority areas for further study. It will provide a valuable resource for researchers, decision-makers and stakeholders in the food industry.

CATHERINE ESNOUF is the Deputy Scientific Director for Food at INRA. She is the Head of a national research network that promotes research partnerships and industrial innovation on the sensory and nutritional quality of food.

MARIE RUSSEL is Research Engineer and a Scientific Project Manager at INRA. She uses interdisciplinary approaches involving food sciences and the humanities to investigate food patterns and preservation issues.

NICOLAS BRICAS is a Researcher at CIRAD who specialises in food security policies and projects. His main areas of research include the effects of urbanisation on food habits in Africa and Asia and the role of small-scale food processing in food security and poverty alleviation.

Food System Sustainability

Insights from duALIne

Edited by

CATHERINE ESNOUF
INRA, France

MARIE RUSSEL
INRA, France

NICOLAS BRICAS
CIRAD, France

CAMBRIDGE
UNIVERSITY PRESS

CAMBRIDGE
UNIVERSITY PRESS

University Printing House, Cambridge CB2 8BS, United Kingdom

One Liberty Plaza, 20th Floor, New York, NY 10006, USA

477 Williamstown Road, Port Melbourne, VIC 3207, Australia

314-321, 3rd Floor, Plot 3, Splendor Forum, Jasola District Centre, New Delhi - 110025, India

103 Penang Road, #05-06/07, Visioncrest Commercial, Singapore 238467

Cambridge University Press is part of the University of Cambridge.

It furthers the University's mission by disseminating knowledge in the pursuit of education, learning and research at the highest international levels of excellence.

www.cambridge.org
Information on this title: www.cambridge.org/9781107036468

First published 2013

A catalogue record for this publication is available from the British Library

Library of Congress Cataloging in Publication data
Food system sustainability : insights from duALIne / edited by Catherine Esnouf, INRA, France, Marie Russel, INRA, France, Nicolas Bricas, CIRAD, France.
    pages   cm
ISBN 978-1-107-03646-8
1. Food industry and trade.   2. Sustainable development.   I. Esnouf, Catherine, 1956– editor of compilation.   II. Russel, Marie, editor of compilation.   III. Bricas, N. (Nicolas), editor of compilation.
HD9000.5.F5995   2013
338.1 9–dc23

2012037614

ISBN 978-1-107-03646-8 Hardback

Contents

Foreword

How will it be possible to feed the world in the demographic context of the twenty-first century?

The diversity of food systems throughout the world is a key element if we are to achieve this.

Little studied in terms of their sustainability, food systems have thus far been considered in terms of the challenges facing agriculture. 'Food sustainability' is unquestionably a major issue for the years to come. From the farm gate to the consumer's plate, these systems now need to be analysed in terms of their contribution towards the pillars of sustainability. However, it is clear that any debate today on this issue cannot be based on consolidated knowledge; research therefore has a major role to play in facing this challenge.

Foresight workshops were initiated in related fields in 2009 and 2010: one on the adaptation of agriculture to climate change (ADAGE) and the other on plant species and sustainable production systems for chemistry and energy (VegA).

Regarding food issues in the longer term, INRA and CIRAD jointly performed a foresight study on the possible futures of farming and food systems worldwide in 2050: the Agrimonde platform thus laid the foundations for agricultural research within a framework of sustainability. In the same vein, our two organisations decided to mobilise private sector and academic experts working in this area to study the trends affecting global food systems in terms of their effects on the environment, health, social equality and the economy.

Launched in November 2009 and completed in June 2011, this project, called duALIne, involved some 125 experts contributing to 10 working groups. Inter-workshop seminars, two general assemblies and a public symposium to debate the preliminary findings marked different stages in the gradual integration of the results. Work in progress was regularly reviewed by a Steering Committee composed of representatives of the different stakeholders (agricultural organisations, industry, the retail industry, environmental protection associations, consumer associations, national and international government representatives and qualified personalities).

We would like here to extend our sincere thanks to all the experts involved and to all members of the Steering Committee, for their participation, their contributions and their advice, all of which enabled us to complete this exercise. Its findings will be used to guide research by our organisations, and we hope that this project will also enable a broader mobilisation of the national and international community to ensure greater food sustainability for future generations.

Marion Guillou
President of INRA
Former President of Agreenium

Gérard Matheron
President of CIRAD

Acknowledgements

The editors of *Food System Sustainability* thank the members of the Monitoring Committee for their involvement and guidance.

They also thank the librarians for their active support, and the authors, contributors and all the experts for their involvement during the 18 months of the duALIne project.

Introduction

Authors: CATHERINE ESNOUF, NICOLAS BRICAS AND
MARIE RUSSEL

Contributor: ARMELLE CHAMPENOIS

Food in the context of sustainable development

Food sustainability must meet the challenges of sustainable development. These challenges, of a political nature, are: (1) the transformation of technical and economic development methods to render them compatible with long-term environmental requirements, (2) intra-generational equity to provide the conditions for joint action to promote essential public goods at different territorial scales up to a planetary level, and (3) the acceptance by today's generations of the costs, duties and limits determined with regard to the fate of future generations, even in the very long term (Godard, 2009). Food sustainability can be defined in this context, and has been assigned numerous definitions. The most recent was proposed by the FAO following a symposium on biodiversity: a sustainable food system will protect and respect biodiversity and ecosystems, be culturally acceptable, economically fair and affordable, nutritionally adequate, safe and healthy, while optimising the use of natural and human resources (FAO, 2010b).

The objective of food sustainability must be to face all or some of these challenges and not to assume that existing models continue to be applied in the long term; this may mean developing new models or adapting existing models.

The food challenge is to ensure that all populations benefit from a diet that meets their qualitative and quantitative needs in a context of sustainable development. The concept of food sustainability includes both an ability to satisfy vital needs and the provision of conditions to drive the global food system towards respect for the three pillars of sustainable development.

Literature reviews and debates among both professionals and the general public have revealed that the notion of sustainability has often been limited in the past: initially, in the 1980s, to impacts on the consumption of fossil fuels (pioneering studies by Pimentel *et al.*, 1973) and, since the 1990s, to impacts on the environment. In the media or in public debates, the expression 'food sustainability' is often restricted to alternative systems, rapidly endowed with a whole host of benefits. Long-term foresight studies on food systems, although they may not use the term 'sustainability', tend to focus on global equilibriums. This challenge is certainly of considerable importance, but it takes no account of the fact that the food system is much more than simply a question of satisfying nutritional needs from quantitative and health points of view; the food system is also profoundly cultural, consumerist, social, economic and local.

One specificity of the field of food is that it lies at the crossroads between individual choices, which determine each diet, and the broad range of socio-economic stakeholders involved, from producers to consumers. The result is that people find it difficult to understand their own, personal impact on the sustainability of food systems.

The objective of duALIne (sustainability of food systems faced with new challenges)

The aim of this project was to review the international literature and mobilise appropriate experts in order to identify major gaps in our knowledge of these challenges, and thus determine priority research areas for the national and international scientific communities.

The experts who took part in this exercise were from academic circles, public authorities, non-governmental organisations and companies.

The workshops were organised and interlinked as shown in Figure I.1. The work of each workshop then generated a chapter in this book. The workshop numbers correspond to their respective chapters.

Scope of the project

The project covered food systems from the farm gate to consumption (cost and availability of products, satisfaction of needs). It did not analyse trends in agricultural production systems as such, but only the results of changes in supply interacting with changes to the composition of diets. In this respect, it was distinctive from, and complementary to, several national and international projects that have analysed the global challenges linked to agriculture (Paillard

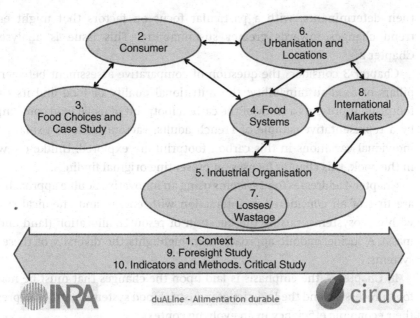

Figure I.1 Organisation of duALIne and of this book.

et al., 2010) or have developed research objectives for sustainable farming (Boiffin *et al.*, 2004; Foresight, 2011; Hubert, 2002; Pretty, 2008; Soussana, 2010).

Nor did this project focus as such on the general challenges linked to the impact of diet on health and well-being, as numerous strategic research initiatives have addressed this issue (Joint Programming Initiative (JPI) 'Healthy diet for a healthy life', French Presidential Plan on Obesity, National Nutrition and Health Programme (PNNS), National Food Programme (PNA), etc.). However, it did examine the interactions between diet and sustainability as potential sources of new questions.

Organisation of this book

The questions for research are designed to clarify the determinants that may support positive or negative elements relative to certain challenges. The project thus involved identifying the major challenges linked to food sustainability, and particularly those specific to this area. This is the subject of the first chapter.

Subsequent chapters analyse the components of food systems in terms of demand, supply and the major issues associated with them.

An analysis of medium and long-term changes in **food consumption** is necessary to address their foreseeable consequences and then consider their determinants, with a particular focus on factors that might enable trend changes towards greater sustainability. This issue is analysed in Chapter 2.

Chapter 3 considers the question of comparative assessment between two pillars of food sustainability: the nutritional quality of food and its carbon footprint. It presents a study of the carbon footprint of usual food consumption by a representative sample of French adults. Factors liable to explain inter-individual variations in this carbon footprint are explored. Unlike elsewhere in the book, this chapter focuses on presenting original findings.

Chapter 4 addresses food systems using an innovative double approach. They are first of all considered in interaction with energy and chemical systems within ecosystems, raising the question of resource allocation (land and bio-mass). A socioeconomic approach then highlights the diversity of these food systems.

In Chapter 5, the emphasis is laid upon the changes that must be made to food processing and the logistics of industrial food systems in order to preserve their economic efficiency in an evolving context.

When dealing with the sustainability of food systems, the spatial dimension merits particular attention. Chapter 6 explores first, how increasing urbanisa-tion questions the sustainability of food supply systems for urban dwellers, and second how the location of different activities in the food chain markedly affects the environmental balance of food systems.

Food losses and wastage, the current lack of knowledge as to their extent, the mechanisms in play and the roles of different stakeholders are so important that Chapter 7 is devoted entirely to this subject.

Food sustainability issues cannot be restricted to the national context, insofar as the international interactions involved are unquestionable. Chapter 8 thus focuses on their role, with particular emphasis on the volatility of global food prices and on public and private standards.

Food systems throughout the world are constantly evolving, even if only certain parameters are considered, such as the trading of foodstuffs, the con-servation techniques employed or the location of retail outlets. Because it is impossible to predict the food systems of the future, Chapter 9 adopts a fore-sight approach to try to comprehend future changes.

Chapter 10 deals specifically with the methods used to assess sustainability, due to their crucial importance, analysing them with respect to their relevance to food and identifying the specific questions that they raise.

1

Context: new challenges for food systems

Authors: CATHERINE ESNOUF AND NICOLAS BRICAS

Contributors: ARMELLE CHAMPENOIS AND MARIE RUSSEL

The global food context is characterised by growing uncertainties and increasing constraints.

The two sides of the global equation for food self-sufficiency are known: the planet will have to provide food for more than 9 billion people in 2050, while satisfying growing non-food demands because of the increasing rarity of fossil energy sources, all in the context of development that is respectful of both the environment and human beings. The joint INRA–CIRAD *Agrimonde* foresight study (Paillard *et al.*, 2010) showed that achieving these ambitions is not impossible if certain conditions are met; in particular, a sustainable increase in yields on the supply side and an improvement in the use of agricultural products at different stages, including a reduction in losses and wastage between the farm gate and the consumer's plate, and possible reductions in, or modifications to, the content of different diets on the demand side. Increased and more secure trade is also necessary, insofar as Africa and Asia, where the greatest population rises are anticipated, will remain deficient in agricultural and food products. Finally, the food context of the future is characterised by increasing uncertainties and considerable constraints.

These different elements provide a framework for the challenges that will have to be met by food sustainability.

1.1 Demographics and uncertainties

Demographic forecasts agree that the global population will start to stabilise around 2050; that is why this time horizon is often chosen as the

Food System Sustainability: Insights from duALIne, eds. Catherine Esnouf, Marie Russel and Nicolas Bricas. Published by Cambridge University Press. © Cambridge University Press 2013.

5

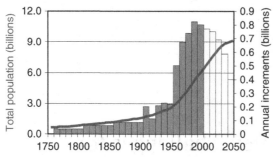

Figure 1.1 World population growth from 3.1 to 6.5 billion people between 1950 and 2000, and towards 9 billion by 2050 (Source: UN, 2003).

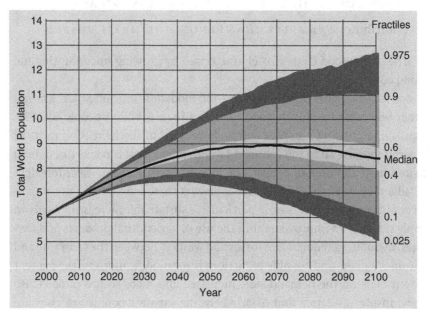

Figure 1.2 Uncertainties regarding population growth (Source: Lutz *et al.*, 2001).

target for foresight studies. The forecasted figure is around 9 billion inhabitants, i.e. a 50% increase over the population in 2000 (6 billion) and a 43% increase over 2005 (Figure 1.1).

However, considerable uncertainties remain, in particular in relation to future birth control patterns. Thus the forecasts range from 7 to 10 billion, although one scenario, with a very low probability of occurrence, situates a possible peak at 12 billion in 2100 (Lutz *et al.*, 2001) (Figure 1.2).

Two major demographic elements are, first of all, the ageing of the population, which will affect both developed and emerging countries, insofar as the average

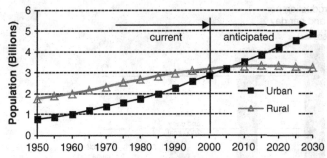

Figure 1.3 In 30 years' time, two-thirds of the world's population will be living in urban areas (Source: UN, 2003).

global age was 28 years in 2005 and should be 38 years in 2050 (Lutz *et al.*, 2008), and second, urbanisation. Although at present half the world's population lives in urban areas, by 2050 this proportion will have reached 70% (Figure 1.3).

1.2   The challenge of global food security

The World Food Summit in 1996 considered that 'food security is assured when all people at all times have economic, social and physical access to sufficient, safe, nutritious food that meets their dietary needs as well as their food preferences and allows them to maintain a healthy and active life' (FAO, 1996).

In quantitative terms, it is important to make a distinction between apparent availability and actual consumption. Apparent availability concerns the quantities produced, to which are added imports and increases in stocks, and from which are deducted exports, reductions in stocks and estimated harvest losses. It does not take account of losses and wastage after harvest (estimated at between 30% and 50% in developed countries). For this reason, it only imperfectly reflects consumption; thus apparent availability in developed countries is about 3500–4000 kcal/day, while actual consumption, closer to nutritional needs, is about 2000 kcal/day. However, figures on availability are the only obtainable data for all countries of the world and for each year since 1960 (Figure 1.4).

It is important to remember that, on average, 57% of initial calories are not consumed.

The current average global availability is 2800 kcal/person/day (Smil, 2000). This is very unevenly distributed, from 2160 in the least advanced countries to 3730 in North America. For this reason, undernourished populations are also very unevenly distributed (Figure 1.5) (FAO-UNO, 2009).

According to the FAO, under-nutrition is when the calorie intake is lower than minimum dietary energy requirements (MDER). The MDER represents the

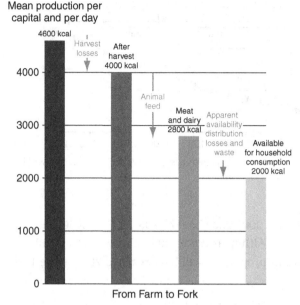

Figure 1.4 From the farm to the fork; the path taken by kilocalories (in kilocalories per day and per person) (Source: Smil, 2000).

amount of energy required to carry out normal activities and have an acceptable minimum weight for a given height. They differ from one country to another and vary each year according to gender and the age pyramid (FAO, 2010c).

The FAO calculation method used in Figure 1.5 takes account of calories. It estimates the number of undernourished people based on the data available on the population (calculation of energy needs by age category: data revised in 2008), food production, trade and the distribution of food products or income. This method has frequently been criticised for its deficiencies and approximations, and other institutions (USDA, World Bank, etc.) use different calculation methods. A recent article referred briefly to these different indices, their use, advantages and drawbacks (Masset, 2011).

Food insecurity is often linked to insufficient agricultural production. However, this production has been growing in different parts of the world, with the notable exception of the former USSR and Africa, where the increase has been lower in absolute terms (Figure 1.6).

Although agricultural production and food availability per capita have increased, the number of people affected by food insecurity started to rise in the mid-1990s, having fallen during the two previous decades (1970s and 1980s) (Figure 1.7).

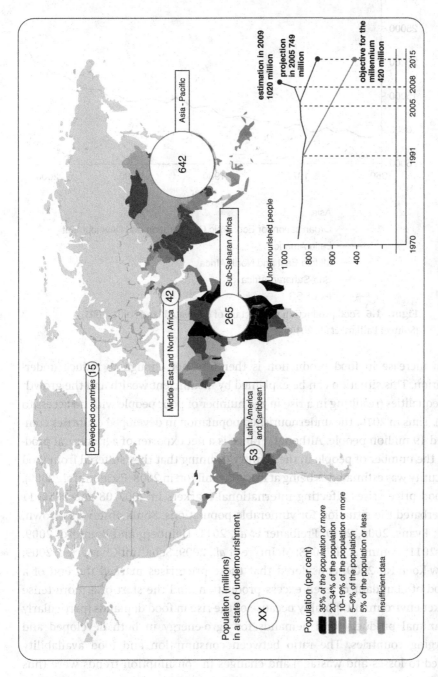

Figure 1.5 Uneven distribution of undernourished populations (Source: FAO-UNO, 2009).

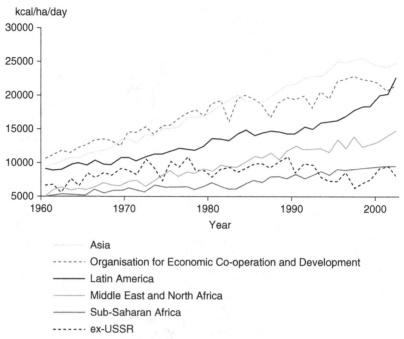

Figure 1.6 Food production per hectare of cultivated land (1961–2003)
(Source: Paillard *et al.*, 2010).

An increase in food production is therefore not enough to reduce under-nutrition. This situation can be explained by insufficient wealth and the growth of inequalities (resulting in a rise in the number of poor people with no access to food). Thus in 2010, the undernourished population in developed countries comprised 19 million people. Although France is a net exporter of agricultural products, the number of people in the country claiming that they suffered from food insecurity was estimated as being at least one million in 2008 (Escalon *et al.*, 2009).

Food price crises affecting international markets in 2007–08 and 2010–11 exacerbated the situation for vulnerable populations. Some observers (Brown, 2011; Evans, 2009, 2010; Freibauer *et al.*, 2011; Heinberg and Bomford, 2009; IMF, 2011; Koning *et al.*, 2008; McIntyre *et al.*, 2009; Schaffnit-Chatterjee, 2009; Sjauw-Koen-Fa, 2009) considered that these price rises marked the end of a period of abundant or even excess production, and the start of a more tense market environment, notably because of the rise in food demands (particularly for animal products) and demands for agro-energy in both developed and emerging countries. The ratio between consumption and food availability (linked to losses and wastage) and changes in consumption trends were thus called into question. The financial, banking and economic crisis of 2008–09 increased the number of undernourished people by 100 million.

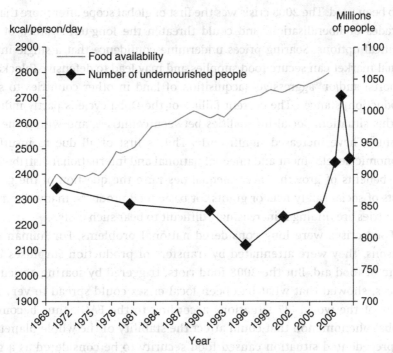

Figure 1.7 Trends in food availability and the number of undernourished people in the world (FAO, 2010).

According to the *Agrimonde* Foresight Study (Paillard *et al.*, 2010), if two consumption scenarios are compared, one pursuing the trend of an increase in calories of animal origin and a trend diet in OECD countries, and the other with a 25% reduction in available calories in OECD countries and the halving of calories of animal origin (from 1200 to 500 kcal/person/day) in the same countries, global agricultural needs for 2050 would fall from 53 000 Gkcal/day to 37 000 Gkcal/day. This would ensure a quantitative equilibrium, without marked increases in yields and cultivated land areas. To take up the challenge of future changes, it therefore appears necessary to question consumption patterns, and no longer simply the means of increasing availability.

The farming and food policies implemented since the 1980s have been marked by a liberal vision: a reduction in support for national agricultural production, the liberalisation of trade to favour comparative advantages and reduce product costs, an extension of this trade to overcome the risks of failure in particular regions. So that consumers can benefit from the lowest prices in the event of local supply deficits, imports and the implementation of safety nets at national and international levels, particularly via food aid mechanisms, have also been used. The 2008 crisis was the first of global scope after more than two decades of liberalisation, and could threaten the long-term future of these political options. Soaring prices undermine confidence that a single international market can secure food supplies and may lead to defensive (blockage of exports) and/or 'aggressive' (acquisition of land in other countries to secure production) stances. The current failure of the Doha cycle is partly indicative of this situation. Social inequalities between countries, and within the same country, have increased significantly. This is first of all due to insufficient economic development and unequal national and international distribution of the benefits of growth. These inequalities raise the question of the growing costs of social safety nets or grants for basic food products; many low-income countries are finding it increasingly difficult to bear such costs.

Food crises were long considered national problems. For humanitarian reasons, they were attenuated by transfers of production surpluses in the form of food aid. But the 2008 food riots, triggered by soaring agricultural prices, showed that what had been local crises could spread to very many parts of the world (37 situations according to the FAO), thus becoming a global phenomenon that could affect the stability of the whole planet. This unprecedented situation caused food security to be considered as a global public good and no longer simply a local problem. However, the methods used to manage such crises, if they were to recur, will probably have to evolve in the context of a possible reduction of agricultural surpluses in countries that were traditionally major exporters. This once again questions the consumption patterns that now affect the international trade balance and food security. Access to food for all countries is becoming a problem for each individual country.

1.3   Climate change

Past trends have demonstrated a 0.8°C rise in mean global temperatures over the past 140 years, and a significant 0.6°C rise over the past 50 years (Figure 1.8) (NASA-Goddard Institute for Space Studies, 2012).

This evolution has not only caused global warming and an impact on agricultural production conditions. Indeed, in its 2007 report, the Intergovernmental Panel on Climate Change (IPCC) noted that one of the consequences of climate change is an increase in the frequency and severity of major climatic events, droughts and floods. This situation may contribute to increasing the instability of production and hence that of markets and prices (Pachauri and Reisinger, 2007).

These changes to the climate are mainly linked to anthropogenic activities and their greenhouse gas (GHG) emissions: farming accounts for 14%, changes in

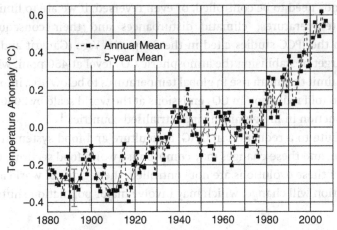

Figure 1.8 Rate of climate change during the past 130 years (Source: NASA-Goddard Institute for Space Studies, 2012).

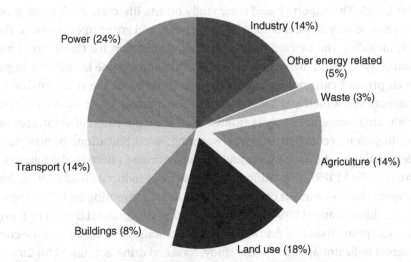

Figure 1.9 The greenhouse gas effect: a natural phenomenon reinforced by human activities (CO_2, CH_4, N_2O, etc.). Total emissions in 2000: 42GtoeCO$_2$ (Source: Pachauri and Reisinger, 2007).

land use 18% and transport 14% (Figure 1.9). Emissions by the food industry are split into different sectors (industry, transport, energy, construction). An approximation of these emissions via energy consumption is shown in Figures 1.10 and 1.11, even though, in the case of agriculture, energy and GHG differ markedly (N_2O, CH_4, soil carbon, etc.).

These trends need to be controlled, or even reversed, if we are to limit their effects on temperatures, climatic disturbances and their consequences. According to the IPCC studies cited in the *Prospective Vega* (Gauvrit and Mora, 2010), the target of stabilising the atmospheric CO_2 level at 450 ppm, which is necessary to limit the mean rise in global temperatures to below 4°C, implies on average a two-fold reduction in CO_2 emissions at the world scale by 2050, and a four-fold reduction in high-emitting industrialised countries.[1]

To contribute to a reduction in GHG, agriculture and food systems require major reform, and these changes may compromise world food security. On the other hand, if these evolutions are not controlled, the regions of world agricultural production will change, which may involve major population shifts.

1.4   Impacts of food systems on the environment

In 2004, the European Commission initiated a study to identify the categories of consumer goods with the greatest environmental impact within the EU-25. The experts based their study on the life cycle analysis of products (goods and services) consumed by households and governments within the EU-25, including the extraction of resources, production for the internal market and waste use and management (both within and outside EU-25). The impact of each product category was characterised in terms of the total volume of the category consumed and then related to the amount spent on it in Euros. The eight environmental impacts studied were the exhaustion of abiotic resources, acidification, ecotoxicity, global warming, eutrophication, human toxicity, destruction of the ozone layer and the formation of photo-oxidant agents. This study, called EIPRO (Environmental Impact of Products) (Tukker *et al.*, 2006a), showed that three major categories of products, accounting for 60% of consumer expenditure, caused 70% to 80% of the total environmental impact of European consumption: food and drink, transport and housing. According to the environmental indicator applied for the study, food and drink accounted for 20% to 30% of global impact, with meat and animal products alone representing between 4% and 12% of the total impact of European consumption.

---

[1] In France, the 'factor 4' concept is used at the national scale to indicate the target of dividing greenhouse gas emissions by four between now and 2050. This 'factor 4' target was successively included in the French national strategy for sustainable development in June 2003, in the 'Climate Plan' in July 2004, and in the 'Programming Law Determining the Orientations of Energy Policies' in July 2005, and was confirmed in 2007 by the French Environment Round Table (*Grenelle de l'Environnement*). 'Factor 4' was initially used by members of the Club of Rome to designate more broadly a multiplication by four of the efficiency of the production modes necessary to preserve resources and energy.

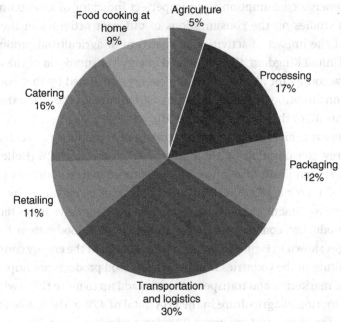

Figure 1.10 Breakdown of energy consumption by different activities in the British food system in 2002 (from Smith *et al.*, 2005).

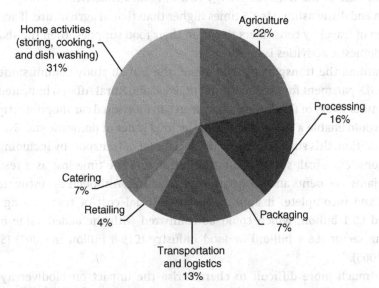

Figure 1.11 Breakdown of energy consumption by different activities in the USA food system in 2000 (Heller and Keoleian, 2000).

Although energy consumption is an imperfect indicator of environmental impacts, two studies on the consumptions of different activities in the food system reflect the impact of activities downstream of agricultural production itself. In the United Kingdom, the cumulated energy consumption of the entire food system accounts for about 13% of the energy consumed by that country. Within this consumption by the food system, agriculture (in the strictest sense) only represents 5% of this energy (Figure 1.10).

The structure is similar in the USA, but agriculture accounts for a much larger share of energy consumption by the entire food system, i.e. 22% (Heller and Keoleian, 2000). Significant divergence in the figures between countries can be seen (a ratio of 1:4 for agriculture). This divergence can be explained by differences in the measurement methods used, and also by the relative importance of agricultural production compared to that of imports into the food system. Indeed, the percentages shown in Figure 1.10 do not take account of the energy consumption of agriculture in the countries from which the food products are imported.

Processing, marketing and transport activities add up to more than twice the energy consumption of agriculture in the USA (total of 47% of the whole energy consumption for those sectors *versus* 22% for agriculture, see Figure 1.11). Overall, 7.3 calories need to be spent within the agricultural and food system to produce 1 dietary calorie in the USA.

In the case of the USA, the energy cost of domestic activities (storage, preparation and dishwashing) is 1.5 times higher than that of agriculture. If account is taken of travel by consumers to obtain their food supplies, then the share of these domestic activities is even larger.

Regarding the transport of food alone, the British study commissioned by DEFRA (Department for the Environment, Food and Rural Affairs) indicated that the home delivery of food products, rather than individual car shopping trips for food, would enable a 70% reduction in the food miles of domestic supply.

In addition, this study evaluated the social cost of transport by including CO_2 emissions, chemical, sound and air pollution and the time lost as a result of traffic jams, accidents and transport infrastructures. Although the estimate was rough and incomplete, it appears that the social cost of transporting food totalled £9.1 billion, which could be compared with the added value of the farming sector (£6.4 billion) or food industry (£19.8 billion in 2002) (Smith *et al.*, 2005).

It is much more difficult to characterise the impact on biodiversity: the pressure of intensive farming on ecosystems has been clearly identified, as has the reduction in genetic resources used for food, despite the diversity of finished products. Agronomic constraints are not the only factor weighing on the reduction in the number of varieties grown; this is also affected by food industry processing requirements, as both processes and equipment are optimised for use of highly specific raw materials. Less well known is the role of biodiversity in providing protection against the spread of pathogens among humans, animals and plants (Keesing *et al.*, 2010).

These different elements thus demonstrate that the environmental impact of food, in particular downstream in the sector after harvest, is a major issue.

1.5   Limited resources: the energy challenge

Data in this section were taken from the *VEGA* Foresight Study (Gauvrit and Mora, 2010).

The total global demand for primary energy has now reached around 12 billion tonnes of oil equivalent (TOE) per year (OECD, 2008a). Energy consumption has risen markedly during recent decades, in terms of both global and individual consumption. The relative shares of different parts of the world are evolving: demand from emerging countries is rising rapidly because of their sustained demographic and economic growth and their current tendency towards energy-intensive growth models.

However, the situation remains highly differentiated between countries and parts of the world. The energy mixes are clearly different, and access to energy extremely contrasted.

The figures on future oil production, and estimates of total global reserves, are the source of major controversy. According to the different analyses, the peak (or ceiling) of oil production is situated between 2015 and 2040.[2]

A drop in oil production before 2050 is thus considered to be highly probable, but in the shorter term, the risks of tension are mainly linked to investment in prospecting, extraction and refining capacities (IEA, 2008), and the ability to mobilise unconventional oil resources. In the short and

---

[2] According to the most pessimistic forecasts, in a little more than a century, half of the conventional oil reserves discovered (or 1000 billion barrels) will already have been consumed. The second half will be consumed much more rapidly than the first, because of much higher global energy demands (according to the ASPO (Association for the Study of Peak Oil)). According to the more optimistic experts, some of the oil that could not be recovered or exploited yesterday is being targeted today and will be much more so tomorrow, because of technological developments. Thus, for example, according to the USGS (United States Geological Survey), more than 2000 billion barrels (mean value) are still to be produced. Furthermore, unconventional oil resources are estimated by the IAE at 7000 billion barrels (in the form of heavy oil, tar sands and bitumous shale), but using current technologies which involve extremely high energy, economic and environmental costs. See report by the Institut Français du Pétrole (IFP) (2012) Quel avenir pour le pétrole (what future for oil?) at: http://www.ifp.fr.

medium terms, the future evolution of oil prices is highly uncertain, but tensions affecting supply will undoubtedly considerably accentuate oil price volatility.

This tension affecting the prices of fossil resources, their instability and their increasing rarity will have a determinant impact on food. Indeed, food is highly dependent on these resources in both industrialised and emerging countries, both for agricultural production and further downstream (transport of products and consumers, the cold chain and refrigeration, the cracking of raw materials, etc.). This foreseeable trend will increase price instabilities, and affect the costs of raw material production, transport and processing, etc.

For all these reasons, current levels of energy use are not sustainable, and as well as diversifying energy sources, it is necessary to reconsider current food systems by working on lower-energy production and consumption systems.

Box 1.1   Pressure on resources and competition for land use for biofuel production

A study has shown that if, using current technologies, all the wheat, rice, maize, sorghum, sugar cane, cassava and sugar beet cultivated today throughout the world were converted into bioethanol only, 57% of global oil demands would be covered (FAO, 2008a; Rajagopal and Zilberman, 2007). This order of magnitude reflects the transformations that might be induced by growth of the biofuels market. Furthermore, the ambitious targets of European Union policies with respect to biofuel development are confronted by this question. Jacquet *et al.* (2007) thus estimated the land area required to satisfy the incorporation target – initially fixed at 5.75% for 2010 – at about 13 million hectares; i.e. slightly less than 20% of currently cultivated arable land in the EU-25 (if production were wholly domestic). The 2007 report by the European Environment Agency estimated that the land available for biofuel production in Europe (without causing environmental damage) could not attain the 10% incorporation target defined for 2020 using the techniques currently available.

There is much debate over land availability and use at the global level: this is partly due to hopeful (or even unrealistic) forecasts regarding an improvement in the energy efficiency of products from biomass. However, this issue has not been covered by duALIne; the current foresight INRA–CIRAD Agrimonde Terra will be focusing on this area.

Furthermore, pressures on natural resources (water, soil) may increase in the future, notably because of greater agricultural intensification. Thus, although increasing quantities of water are being extracted and numerous parts of the world are experiencing supply problems, future food demands will increase considerably and competition will grow between agricultural, domestic and industrial uses for water, in addition to the anticipated effects of climate change. In regions with scarce or already insufficient water resources, the development of supernumerary irrigated crops for non-food uses is likely to accentuate these problems (de Fraiture *et al.*, 2008; Fischer *et al.*, 2009). The quality of water resources will also be affected if biomass production causes an increase in the use of chemical inputs (Lorne and Bonnet, 2009) and if processing-related pollution is not controlled (e.g. effluents containing high levels of organic matter).

Finally, depending on the systems implemented and the regions concerned, a major development of new crops may participate in accelerating soil degradation, a phenomenon that has become crucial at the global scale. In this respect, the possibility of using perennial plants, notably for lignocellulosic bioproducts, may however promote soil protection. In addition, the use of crop residues – a subject of growing interest – requires considerable uptake control to maintain sufficient levels of organic matter and preserve soil structure and fertility.

1.6   Nutritional trends (accelerated in emerging countries)

For more than 50 years now, studies performed using FAO data have enabled the characterisation of the nutritional trends that accompany economic development (Cépède and Lengellé, 1953; Cépède and Lengellé, 1970). These structural changes to the diet (reduction in the proportion of carbohydrates, increase in lipids, stability of protein calories) are directly linked to increases in the consumption of animal products with a rise in income (Périssé *et al.*, 1969). Since the early 1960s, this trend has been confirmed in developed countries and is now gradually spreading in emerging countries, where it is characterised by a very rapid transition rate: in 40 years, the emerging countries have caught up with developed countries. However, this trend raises problems regarding both resource use (energy, land, water, etc.) and health. We shall be discussing its impact on obesity below.

1.7   The health challenge

The development of obesity and the growth of resulting health inequalities are now essential factors. Obesity is defined by a body mass index (BMI:
body weight divided by the height squared) higher than 30.

Epidemiological data are becoming concerning: the World Health
Organization (WHO) estimates that 400 million people are currently affected
by obesity throughout the world, i.e. 7% of the global population; this figure may
have reached 12% by 2020 if current trends continue. More recent figures
(Finucane *et al.*, 2011) give a global estimate of 500 million obese subjects. The
worldwide rise in body mass index is now 0.4 kg/m^2 per decade among men and
0.5 kg/m^2 among women. In some countries, obesity rates rose considerably
between 1980 and 2008; for example, from 12% to 30% in North America (men)
or 20% to 37% in South Africa (women) (Figure 1.12).

In France, the ObEpi/INSERM studies (ObEpi, 2009) showed that obesity
affected 8.7% of the population in 1997 and 14.5% in 2009. A study by the
IRDES (Sermet, 2006) in France on the data for 2002 placed the cost of obesity at

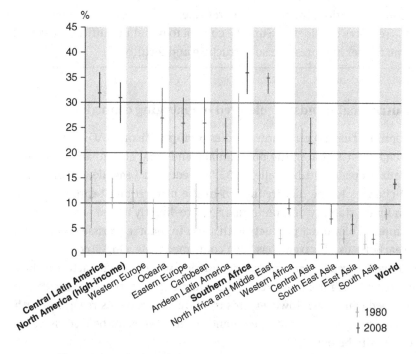

Figure 1.12 Rise in BMI among women between 1980 and 2008
(Source: Finucane *et al.*, 2011).

between 1.5% and 4.6% of overall healthcare expenditure, or €2.6 billion per year, of which €2.1 billion were paid by health insurance schemes. Obesity plays a major part in the development of a whole series of chronic diseases, including non-insulin-dependent diabetes (more than 80% of cases are obesity-related), hypertension, cardiovascular diseases and also certain cancers, respiratory diseases and joint diseases. The mean consumption of medical care and goods by an obese person reaches about €2500, double that of a non-obese individual.

More worrying is the fact that this obesity is increasing rapidly in developing countries; it now affects 200 million people, about the same number as those with insufficient body weight (Kelly et al., 2008; WHO, 2009). Some countries are faced with a considerable rise, such as those in Oceania (from 8% to 28% between 1980 and 2008 among women), Central America (from 10% to 33% among women), North Africa and the Middle East (from 15% to 35% among women) (Finucane et al., 2011). This concomitant presence of low weight and obesity can exist within the same household, with obese parents having underweight children (a double burden). It is particularly worrying because research has shown that the environment during a child's development can affect his or her phenotype at adulthood, particularly regarding epigenetic markers. For example, a low body weight during childhood can lead to major adiposity in adulthood, but this will depend on the subsequent nutritional environment (Gluckman et al., 2009).

Obesity is correlated first to social inequalities in OECD countries, with the level of education being a determining factor among women (Mackenbach et al., 2008), and second to the level of income: according to the ObEpi survey in France, the rate of obesity reaches 19% in households with an income lower than €900, falling steadily as income rises, to only 5% in households with an income higher than €5300. This inequality has increased, the difference in prevalence between the two extremes rising from 8% to 14% between 1997 and 2006 (Charles et al., 2008).

Despite the rapid development of diet-related non-transmissible diseases, we must not forget that the problem of nutrient deficiencies has not yet been resolved. First, this concerns protein and energy deficiency, one of the factors underlying growth retardation in children. Height retardation in children below the age of 5 years (stunting) ranges from 35% to 55% in different sub-Saharan African countries, and affects around 45% in India or Bangladesh (Figure 1.13). It affects nearly 180 million children below the age of 5 years throughout the world, i.e. one in three children.

Second, malnutrition concerns micronutrient deficiencies, referred to as 'hidden hunger'. According to WHO (de Benoist et al., 2008), nutritional anaemia

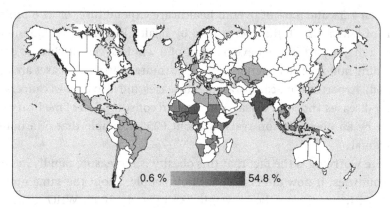

Figure 1.13 Global prevalence of malnutrition in 2007–11: percentage of children below the age of 5 years with growth retardation in relation to global standards (stunting) (Source: World Bank, 2011).

affects more than 2 billion people throughout the world. Half of all pre-school age children and pregnant women suffer from iron-deficiency anaemia. Iodine deficiency affects a billion people and vitamin A deficiency 190 million people.

Insufficient food consumption is not the only determinant of malnutrition. Health and hygiene are two other fundamental factors contributing to nutritional diseases such as childhood growth retardation. In this respect, the safety of food plays a crucial role. Although it has improved considerably in industrialised countries, it remains a serious problem in developing countries. Indeed, in the latter case, and more than in industrialised countries, new problems are appearing that are linked to insufficient controls and regulations on the use of certain inputs such as pesticides. Food poisoning due to melamine, borax or pesticides is becoming increasingly common in these countries.

Malnutrition is not just a consequence of poverty, it is also one of its causes. The economic cost of malnutrition has been estimated by the World Bank at between 2% and 3% of GDP. Inversely, R. W. Fogel estimated that an improvement in nutritional status was responsible for about 30% of the growth in per capita income in the UK between 1790 and 1980 (Fogel, 1994).

Finally, changes in food consumption, and notably the diversification of baby diets at too early an age and the increasing consumption of exotic products or industrial food additives, are suspected as being the cause of food allergies (Wang and Sampson, 2011). It is difficult to quantify their prevalence. Clinically proven food allergies are estimated to affect between 1% and 5% of the total population in different European countries. However, the proportion of people who consider that they have one or more food allergies is markedly higher (Moneret-Vautrin, 2008). Similarly, although improvements in the safety

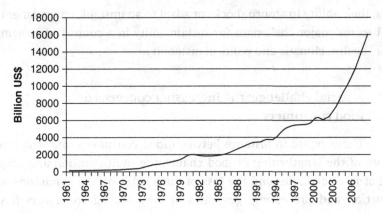

Figure 1.14 Global agricultural exports (FAOSTAT data).

of food have enabled a marked reduction in cases of food poisoning, it can impair the immune defences and increase the population's vulnerability. This issue is complex and controversial (Mannie, 2010; Okada *et al.*, 2010).

1.8   Increased circulation of goods and people

International trade in agricultural products developed considerably during the twentieth century and has seen a further, marked acceleration since 2000. Thus, during the 10-year period between 1998 and 2008, the real value of global agricultural exports increased three-fold (Figure 1.14).

One important consequence of this increase in trade is the development of the circulation of pathogens that affect plants, animals and humans. These exchanges are not the only cause, but they are associated with other determinants such as imbalances affecting ecosystems, the anthropisation of environments and selection pressures exerted on different pathogens.

1.9   Sustainability of the industrial economic model: the vulnerability of low stock level systems

One of the many causes of the 2008 prices crisis was the low global level of stocks, which no longer provided a buffer against a supply deficit. The switch towards just-in-time systems and an acceleration in trade have led to greater risks of shortages and in particular have exacerbated their consequences.

An analysis of different food system management scenarios in the event of supply crises or major epidemics revealed the considerable vulnerability of just-in-time systems (Gilbert, 2007). The resilience or viability of systems (in other words, their ability to absorb shocks or adapt to an unstable environment) have thus become major challenges for sustainability in a context of growing economic, health, climatic and political instability.

1.10   A social challenge: the increasing concerns of food consumers

The increase in distance between food consumers and what they eat because of the lengthening of food chains, the multiplication and empowerment of intermediaries and the loss of knowledge on the conditions of food production and processing, generates anxiety among consumers (Fischler, 1990). They question the sources of products, fearing that their diet is becoming artificial, although they would like it to remain 'natural' and 'authentic', and they find it difficult to accept the contradiction between the globalisation of trade and the need to retain regional roots (Appadurai, 1986). Through its marketing activities, the food sector has indeed understood these anxieties and taken account of criticisms (Boltanski and Chiapello, 1999), proposing 'labelled' products that highlight the values sought by consumers: regional products with controlled origins, fair trade systems involving small producers or organic farmers, etc. The development of alternative short chains and the maintenance of self-sufficiency participate in this search for a new relationship with food. Yet this response through the commodification of alternatives appears insufficient. In France, 40% of the people interviewed for the Nutrition and Health Panel in 2008 indicated that they were not satisfied with the quality of their foodstuffs (Escalon et al., 2009).

More and more, consumers are forced to delegate control over the safety of food to operators in the chain. However, constructing these trusting relationships is no simple task. The learning process is further hampered by scandals linked to industrial negligence or abuse, which foster widespread concern (Apfelbaum, 1998). The consequence is a risk of hypersensitivity to rumours and of erratic reactions by consumers in the event of any doubts, which can lead to harmful consumption behaviour. Such situations also tend to increase market instability for industry (Chevassus-au-Louis, 2002). In the extreme case of eating disorders (anorexia, bulimia), the stigmatisation of obesity and the high value placed on slimness appear to contribute to their development. These conditions do not only have a psychological trigger, which requires individual therapy, they are also the result of dietary modernity and the over-medicalisation of foods (Fischler and Masson, 2008; Poulain, 2009).

1.11   Summary

Historically, in France, issues of sustainability in agriculture (in its broadest sense) were first raised at the production level, and more specifically with respect to farming. The environmental impacts of the intensive production systems that developed during the second half of the twentieth century led to the development of concepts such as agro-ecology, and agricultural models such as organic farming.

Paradoxically, while the food sector – processing, marketing, catering and consumption – appears to pose as many, if not more, problems in terms of sustainability as the farming sector, it is rarely found in research programmes on sustainability. Thus, as shown by this brief review of the context, it seems legitimate to question the tertiarised agro-industrial model (Rastoin *et al.*, 2010). And indeed, its sustainability raises questions in several areas.

In terms of the environment, the food industry contributes significantly to pollution and greenhouse gas emissions, and threatens biodiversity; climate change may contribute to increasing the instability of production and hence that of markets and prices. The increasing scarcity of fossil fuels and increased competition between food and energy uses for agricultural products may also contribute to market tensions.

In economic terms, it seems that the sector will have to develop in an environment that is more unstable than has been seen during the past three decades. This may have potentially important consequences in terms of employment. The current food system has not succeeded in reducing under-nutrition, a problem that has been resurgent during the past 15 years. Tensions in agricultural raw material markets raise fears that prices will remain higher than during the past 30 years, or even soar, if supply deficits persist. Oil prices are also expected to rise and flare-ups are likely in view of the growing tensions in the energy market. The greater circulation of goods at a global scale is likely to increase the risk of safety crises.

In social terms, the growth of inequalities poses not only moral questions but also threatens world stability. The increasing distance between consumers and their food generates anxiety and dietary behavioural disorders.

In terms of health, the issue of food safety is far from being resolved in Southern countries and could generate new problems if the safety boundaries were pushed too far. But it is principally from a nutritional point of view that the industrialised food system is showing its limitations. Diet-related non-transmissible diseases now constitute a major public health challenge.

The uncertainties and risks that weigh upon the future of the food industry raise two main questions: the first concerns the conditions required to ensure that the sector is more able to face up to an unstable and uncertain environment. How can its resilience or viability be enhanced? The second problem is how to reduce tensions in the markets for both agricultural raw materials and energy. Availability must be improved by increasing production and reducing losses, and also to dampen any excessive rise in demand.

These environmental and health challenges have been known for many years and are now on the international political agenda. The economic and social challenges have become more crucial since the 2008 crisis. Economic vulnerability and uncertainty dominate the vision of more industrialised countries. How can research help to better characterise situations, identify their determinants and assess opportunities for action? These are the questions that led INRA and CIRAD to propose a research strategy with respect to sustainable food systems.

2

Consumption and consumers

Authors: PIERRE COMBRIS, BERNARD MAIRE AND
VINCENT RÉQUILLART

Contributors: FRANCE CAILLAVET, ARMELLE CHAMPENOIS,
SANDRINE DURY AND SÉVERINE GOJARD

Analysis of the medium- and long-term evolution of global food consumption highlights some striking regularities. This chapter addresses the
potential consequences of a generalisation of these trends, before focusing on
their determinants and in particular the factors that might alter current trends
to promote greater sustainability. Six main areas are covered: the characteristics
of long-term changes to diet, the international convergence of food models and
evolution of the situation in Southern countries, the bio-physiological determinants of changes to consumption, the role of agricultural policies, the identification of factors that may change trends and, finally, the heterogeneity of
consumption and nutritional inequalities.

One of the major issues addressed by this working group was to determine
whether dietary trends and their underlying major determinants were the same
in all countries, or whether on the contrary it was possible to identify differences that might indicate alternative pathways for change. The same concern
led the group to look at break points in past consumption trends, concerning
meat in particular, insofar as they might lead to a clearer understanding of the
mechanisms underlying the alteration of long-term trends and the emergence
of new dietary standards.

2.1   Characteristics of long-term dietary trends

Work on the history of economics (Bairoch, 1997; Braudel, 1979;
Toutain, 1971), of agriculture (Mazoyer and Roudart, 2002) and of food

Food System Sustainability: Insights from duALIne, eds. Catherine Esnouf, Marie Russel and
Nicolas Bricas. Published by Cambridge University Press. © Cambridge University Press 2013.

economics (Cépède and Lengellé, 1953) has shown that diets evolve under the effect of powerful nutritional and economic determinism. Analysed at a highly aggregated level (macronutrients, food groups), the evolutions have been very similar from one country to another and directly dependent on the level of economic development.

Observed over long periods or vast geographical areas, the effects of economic factors on diet have been spectacular. In developed countries, and now in most countries of the world, the agricultural revolution, sustained and then extended by the industrial revolution, considerably reduced the cost of dietary calories. The relative prices of different foods have been completely disrupted, as have diets, with a marked improvement in the sanitary status of foods. The positive consequences of these changes are numerous, whether with respect to the development of biological potential, aptitude for work, longevity or quality of life (Fogel, 1994). The negative effects, proven and foreseeable, are no less significant (increase in weight and development of obesity, diabetes, etc.). These are now becoming major public health problems, the economic consequences of which should not be underestimated, especially in developing countries (Drewnowski and Popkin, 1997; Schmidhuber and Shetty, 2005), and particularly since this economic trend has concurrently contributed to a significant reduction in food energy needs (less physical activity), a phenomenon amplified by changes to the structure of employment (primary/secondary versus tertiary) and urbanisation.

Thanks to the consumption series reconstituted and analysed by historians (Fogel, 1994; Toutain, 1971), we can now gain a fairly clear idea of the characteristics of dietary changes in Europe since the end of the eighteenth century. The principal stages in this evolution were identical in most countries, even though the phases differed according to specific national events.

In France, for example, this evolution occurred in two phases. The first stage was the agricultural revolution, which was contemporary with the industrial revolution. It was characterised by a very marked increase in the per capita calorie intake during the nineteenth century. Throughout this period, the increase in total consumption resulted from a proportional rise in the consumption of all foods. The least expensive foods (cereals, staple foods) constituted the foundations of the diet, so that during the years 1880–90, when caloric saturation was reached, cereals (mainly in the form of bread) still represented most of the ration (Figure 2.1). During this period, physiological energy needs still remained high, and this evolution mainly had favourable effects on health.

A new phase then started, the 'nutritional transition' itself, characterised by a radical change to the structure of the diet. The consumption of basic foods

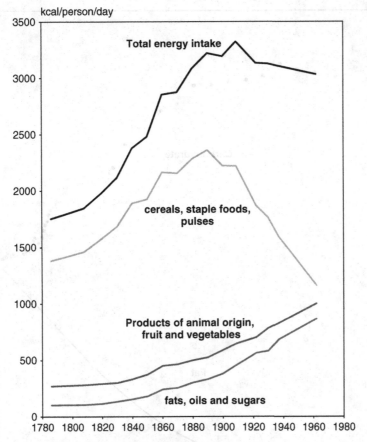

Figure 2.1 Long-term evolution of energy intakes in France (Source: P. Combris, from Toutain, 1971).

(cereals, staple foods, pulses) declined irrevocably, while that of other products (those of animal origin, fruit and vegetables, fats and sugars) continued to increase. Although throughout the quantitative growth phase the nutritional structure of the diet remained almost stable (Figure 2.2), it was very profoundly modified as soon as caloric saturation was attained: between 1880 and 1980, the share of carbohydrate calories fell from 70% to 45% of total energy intake, while the share of calories of lipid origin rose considerably from 16% of energy intake to 42% (Combris, 2006).

This transition process was completed in about 1985–90, from which point the evolution of the relative shares of macronutrients in total energy intake began to stabilise (Figure 2.3). This stabilisation does not mean that the diet is no longer changing. It reflects the fact that the major trend towards replacing basic

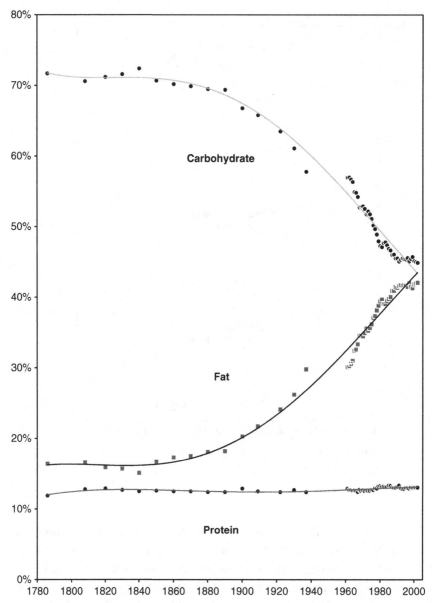

Figure 2.2 Long-term evolution of the structure of energy intakes in France (Source: P. Combris, from Toutain, 1971. Food supply data from FAOSTAT (http://faostat.fao.org/site/609/default.aspx)).

products with meats, dairy products, fats and sugars, has now reached its term. In other words, the saturation of consumption which stabilised the global calorie level at the end of the nineteenth century now concerns all food groups (Combris, 2006).

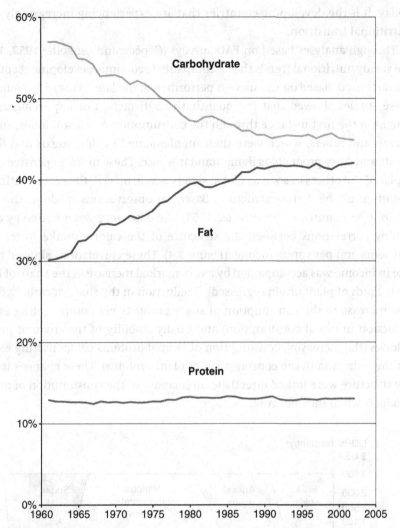

Figure 2.3 Evolution of the structure of energy intakes in France since 1961 (Source: P. Combris, 2006. Food supply data from FAOSTAT (http://faostat.fao.org/site/609/default.aspx)).

2.2   Convergence of food models

2.2.1   *Consumption of animal calories*

The trend described above was of course not specific to France. Whether it started at a very early stage, as in the UK, or slightly later, as in southern European countries, this nutritional transition was completed during the second half of the twentieth century in most developed countries.

Today, it is the developing countries that are experiencing increasingly rapid nutritional transition.

Through analyses based on FAO surveys (Cépède and Lengellé, 1953, 1970), the steady nutritional trends that accompanied economic development could be characterised. Based on the surveys performed in the late 1930s in 70 countries, these studies showed that the quantitative satisfaction of requirements was sought in the first instance through the consumption of 'cheap' foods, such as cereals and tubers, which were then supplemented by fats, sugars and finally meats and dairy products as living standards rose. These more 'expensive' foods replaced the others as soon as global satiety was achieved, thus accelerating the evolution of the dietary structure. Based on observations made in the early 1960s in 85 countries (Périssé et al., 1969), the data were systematised by establishing correlations between the structure of the calorie intake in terms of nutrients and per capita income (Figure 2.4). These correlations showed that a rise in income was accompanied by a very marked increase in the share of lipids (only lipids of plant origin regressed), a reduction in the share of carbohydrates (the increase in the consumption of sugar products not compensating for the reduction in cereal consumption) and finally stability of the share of protein calories (the increasing consumption of animal proteins compensating exactly for the reduction in the consumption of plant proteins). These changes to dietary structure were linked directly to an increase in the consumption of animal products when incomes rose.

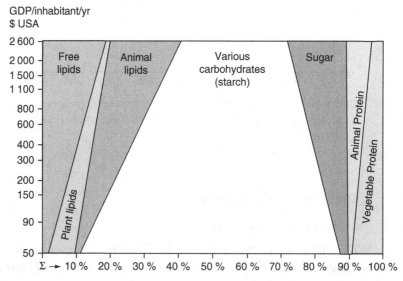

Figure 2.4 Dietary structure by income level worldwide in 1962 (Source: Périssé et al., 1969).

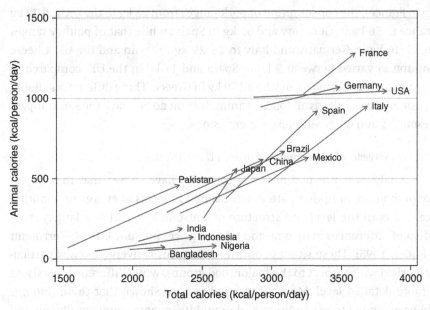

Figure 2.5 The consumption of animal calories and total calories between 1961 and 2005: a few examples from throughout the world (Source: P. Combris, 2006. Food supply data from FAOSTAT (http://faostat.fao.org/site/609/default.aspx)).

Since the early 1960s, this trend has been confirmed in developed countries and has gradually spread to emerging countries (Figure 2.5). Even if, at this highly aggregated level, differences can be seen between countries, particularly those whose dietary customs still involve a large share of plant products (India, Japan, etc.), the consumption of animal calories, evaluated in terms of availability, has risen sharply. In the long term, this increase in the consumption of animal calories in the great majority of countries of the world (the countries concerned accounting for more than 5.3 billion people in 2005) raises the major problems of pressure on agricultural resources and the emission of greenhouse gases (see Chapter 3 and the *Agrimonde* Foresight Study; Paillard *et al.*, 2010).

The generalisation of a high consumption of animal calories does not necessarily mean that food models will become uniform. When the analysis moves away from nutrients and major food groups to focus in more detail on specific products, significant and persistent differences can be seen between countries that are otherwise relatively similar in terms of their economic development. This is the case in Europe, for example, regarding meat and dairy products. Between the major western European countries, the apparent consumption of beef can vary two-fold (13 kg/person/year in Germany, compared to more than

26 kg in France), the consumption of pork ranges from 28 kg in the UK and 32 kg in France to 56 kg in Germany and 62 kg in Spain, while that of poultry ranges from 15–16 kg in Germany and Italy to 28–29 kg in Spain and the UK. Cheese consumption varied between 9 kg in Spain and 11 kg in the UK, compared to 24 kg in France, 25 kg in Denmark and 30 kg in Greece. These differences should be considered carefully as all foods of animal origin do not have the same impact on resources and on greenhouse gas emissions.

2.2.2 Convergence of expenditure and product characteristics

Analyses performed in OECD countries have shown that the convergence of food consumption patterns, measured in terms of energy or quantity, concerned both the level and structure of consumption and was largely independent of differences in income and relative prices (Blanford, 1984; Herrmann and Röder, 1995). These studies confirmed that this convergence was particularly marked with respect to the major food groups, while differences persisted at a more detailed level. More recent studies have shown that these findings could be extended to expenditure and to middle income countries (Regmi and Unnevehr, 2006; Regmi et al., 2008). The latter study involved 47 countries grouped by level of income (see Table 2.1) and covered the period from 1990 to 2004. It concluded that expenditure in major food groups (meats, vegetables, sweet products, soft drinks) converged towards the level of high-income countries, although this convergence also concerned upper middle income and lower middle-income countries (see Table 2.1). Other indicators relative to product

Table 2.1 *Countries included in the analysis of convergence (Regmi et al., 2008).*

High income	High income	High income	Upper middle	Lower middle
Canada	Belgium	Norway	Czech Republic	Brazil
USA	Finland	Switzerland	Hungary	Colombia
Australia	Greece	Singapore	Poland	Peru
Japan	Italy	South Korea	Chile	China
France	Spain	Taiwan	Mexico	Indonesia
UK	Sweden	New Zealand	Malaysia	Philippines
Germany	Denmark	Israel	South Africa	Thailand
Netherlands	Ireland	Kuwait		Algeria
Austria	Portugal	Saudi Arabia		Egypt
		United Arab Emirates		Jordan
				Morocco
				Tunisia

attributes, claims shown on packaging and the characteristics of distribution and catering systems, showed the same trend towards convergence.

2.2.3   Consumption trends in Southern countries

During the 1970s, the major problem faced by Southern countries was one of food insufficiency. In these countries, the transition towards satisfying total caloric needs is not wholly complete, but major progress has been made. At an aggregate level, the FAO estimated that the mean energy intake per person had risen by about 31% in Southern countries (versus an average of 16% world-wide) within a few decades. Thus of the seven Southern countries with more than 100 million inhabitants (China, Indonesia, Brazil, India, Pakistan, Nigeria and Bangladesh), only Bangladesh continued to show very low consumption levels. Today, there are still about 30 countries, a great many of them in sub-Saharan Africa, where the average apparent food consumption per person is lower than 2200 kcal. Forecasts suggest a continuing rise in demand and food consumption, so that the mean for developing countries will now be close to 2850 kcal in 2015 and 3000 kcal in 2030 (Bruinsma, 2003).

This trend has been associated with a clear reduction in the prevalence of the different types of malnutrition (UN-SCN, 2009). Yet, meanwhile, or within a very short time frame, and unlike what has happened in industrialised countries (where the phenomenon has developed over more than a century, as seen above), a new dietary transition has emerged, which first affected rapidly growing emerging countries (e.g. Brazil, China, South Africa) and the wealthier segments of the population (notably urban), before gradually spreading to the entire pop-ulation (Popkin, 2006). This evolution was characterised first of all by a phase involving an increase in global consumption (energy) and then a 'Westernisation' of diets, although these differed between regions and societies and are still poorly understood (Aounallah-Skhiri et al., 2011; Delisle, 2010; Flores et al., 2010; Wang et al., 2008), and undoubtedly involved generational effects accompanying the shift from one to another. Although characterised by beneficial diversification, this trend has favoured greater energy density, with dishes involving increasing quantities of lipids, simple sugars and salt, while fibre and micronutrient require-ments are not always met satisfactorily (Popkin, 2006), thus moving closer to an increasingly generalised global situation (Popkin, 2011).

These dietary changes, accompanied by modifications to consumption behaviour (Wang et al., 2008) and a reduction in the level of everyday physical exercise, have favoured the rapid onset of obesity and chronic diet-related diseases, even though deficiency malnutrition has not completely disappeared. FAO estimates have revealed a marked rise in the number of undernourished people since the beginning of the twenty-first century, with a peak of more than

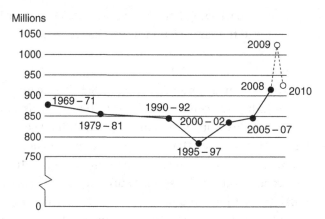

Figure 2.6 Numbers of undernourished people in the world between 1969 and 2010 (Source: FAO).

a billion in 2009 following the sharp rise in food prices and the global economic crisis (Figure 2.6).

However, these estimates are still a matter of debate, as other models, in contrast, have suggested a reduction between 2005 and 2008 (Headey, 2011). These dietary changes have been accelerated by the convergence of new demands from consumers in Southern countries, notably accompanying lifestyle changes linked to urbanisation. They are also amplified by the (limited) positive evolution of incomes, the reduction trend in the prices of certain energy-rich foods (plant oils, animal products), and poor targeting of the aid granted to the agricultural and food sectors in a certain number of these countries (Webb and Block, 2012). Furthermore, the increased firepower (during the past 20 years) of major food and particularly retail groups, linked to the liberalisation of world trade, has also contributed to these changes (Thow and Hawkes, 2009). Yet studies are starting to suggest that although trends are converging between Northern and Southern countries, the same policies to improve diet, health and the environment will not necessarily have the same effects or the same economic and social costs. Research is required to conceive holistic policies adapted to what remain to be rather distinctive environments (Lock *et al.*, 2010).

2.3   Bio-physiological bases for the convergence of food models

Although food consumption continues to vary according to environments and populations, a certain number of physiological elements contribute to a relative convergence of food models, at least in terms of balancing the principal nutrients. These elements of convergence are linked to the everyday need to eat appropriately in terms of quantity and quality and to the regulation of this intake via numerous physical, hormonal and microbiological factors or sensory perceptions; these are also influenced by various psychological or social factors.

2.3.1   Food requirements

Physiological food requirements are defined not in terms of foods but in terms of energy, water and nutrients. Indeed, there is a very broad range of food combinations that can meet these basic needs. There is almost general agreement today regarding minimum energy requirements. Generally speaking, spontaneous consumption is always higher than this level when sufficient quantities are available. Requirements vary according to age, gender, height, corpulence, physical activity and physiological status. Global guidelines have been established by expert committees for populations in good health and leading a 'normal' lifestyle (FAO, 2001). For example, these requirements are estimated at 2450–3450 kcal/day on average for a 30-year old man weighing 70 kg, and at 1750–2300 kcal/day for a 30-year old woman, depending upon whether they have a high or low level of physical exercise. The guidelines laid down by different countries differ in order to take account of the average consumption levels observed spontaneously in the population (for France, the ANSES proposes an average of 2700 kcal/day for men and 2200 kcal/day for women with a moderate level of physical exercise).

We also have quite an accurate idea of the range of macronutrients, lipids and protein, as a percentage of energy needed to meet minimal nutritional requirements. Once again, expert guidelines are available, based on growing consensus. However, they continue to vary between countries, and it is sometimes difficult to relate nutrients to diet-based guidelines. While numerous mechanisms regulate the balance of body carbohydrates and proteins, there is no strong regulation of any imbalance in lipids, and in practice it is quite easy to exceed the recommended levels.

Regarding micronutrients, the minimum level required is also known, but its regulation differs markedly according to the different vitamins or minerals concerned. A large proportion of populations throughout the world fall below the requirement levels for certain micronutrients, partly because their diets are too heavily dependent on plant products and in particular because they are not sufficiently varied.

2.3.2   Regulatory factors

It is a sensation of hunger (empty stomach, intestinal motility, rapid variation in glycogen stocks, hypoglycaemia, etc.) that pushes an individual to seek caloric foods; inversely, a sensation of satiety blocks this need to eat.

Satiety is high with proteins, low with lipids and intermediate with carbohydrates. It is subject to complex regulation via signals from the digestive tract to the central nervous system; it is positively linked to the fibre and water contents of foods, which favour significant stomach filling with a lower quantity of energy, and is generally inversely linked to the energy density of the intake consumed. A certain number of studies have indeed suggested that the calories ingested from drinks may induce less satiety than the same quantity of calories ingested in a solid form. A varied diet promotes an increase in consumption and is able to modify the energy balance over the short or medium term.

The overall energy balance which most individuals achieve normally over weeks or months despite major variations in the amounts of energy ingested or expended (up to ±23% from one day to the next for energy intake) involves complex regulation that is adapted to a given body composition in a given environment, rather than to a fixed body weight; however, an 'error' of just 1–2% in this adjustment can lead to significant cumulative effects over the longer term. Genetics or the early acquisition of mechanisms involving a different management of the respiratory quotient (reflecting the preferential use of a particular 'fuel', carbohydrates or lipids) will naturally influence the evolution of consumption. However, environmental or individual factors are also likely to affect consumption. A greater appetence for particular foods and dishes, a higher proportion of lipids and simple sugars in the diet, changes in energy density (sometimes unknown to the consumer), greater choice, permanent availability, larger portion sizes, improved purchasing power and lower levels of physical activity are all factors that are compatible with an increase in energy intake and a higher accumulation of body fat (Flatt, 2011). Furthermore, conscious mechanisms linked to a personal cultural vision (body image) or broader societal influence do not appear to be able to easily compensate for this trend towards an increasing energy intake. It is as if evolution had promoted the control of energy deficiency to a greater extent than that of excessive energy intake. The influence on total consumption of different macronutrients or of energy density, and the mechanisms in play (appetence versus satiety) have given rise to contradictory findings that require further research (Sorensen et al., 2003). Yet there is little doubt that an energy-dense diet, rich in simple sugars and lipids, causes weight gain.

Furthermore, the hypothesis advanced by Barker regarding foetal metabolic programming may reflect the acceleration in the spread of obesity and its associated diseases in Southern populations. Confronted during infancy with malnutrition and during adulthood with a greater abundance of food, resulting from sustained metabolic reorientation in utero or during childhood aimed at better absorbing and utilising energy substrates, these populations are now finding themselves in some way at odds, equipped with a factor that is favourable in a situation of malnutrition but not in the case of abundance (Barker, 1998).

2.3.3   The role of taste

It is generally considered that taste is an important determinant of food choices. Taste may help in selecting the necessary variety of nutrients. However all human beings do not perceive tastes in the same way. Different densities of taste buds, genetic differences in taste receptors, the sensitivity of these receptors, or components in saliva, all make a notable contribution to these differences in perception and food preferences. Five primary tastes can be distinguished: sweet, salty, bitter, acid and umami. A preference for sweet foods and the rejection of bitterness appear to be present from birth; they have been selected because of their ability to encourage the search for energy in the former case, and to avoid toxic substances in the latter. The appreciation of sweet taste is partly linked to one or more genes, and variations in the perception of sweetness can influence dietary preferences. However, animals that are genetically selected for their sweet taste receptors express varied consumption behaviours. There is therefore another regulation that acts at cerebral level, which would considerably promote the influence of environmental and cultural factors. The perception of umami (monosodium glutamate) is also associated with genes, but the relationships with obesity are contradictory. The appreciation of acidic taste may also be strongly dependent on genetics, but this has not been the subject of much research to date; it appears that it is subject to little variation.

A taste for salty foods is not innate. It appears quite soon after birth, but it seems that there is a relatively low physiological need in humans, unlike animals; salt is therefore consumed out of habit (salting is a useful preservation process) or for pleasure; this hedonic response results from a mixture of biological (hormones, genetics), cultural and environmental factors. It is able to compensate for greater sensitivity to bitterness (Hayes *et al.*, 2010). The appreciation of bitter taste is one of the most widely studied, notably relative to its genetic influences; there are variants linked to greater sensitivity (via 25 different taste receptors which considerably refine this perception, probably linked to the detection of food toxins) and other 'adaptive compensation' variants that limit sensitivity to bitterness, thus allowing a certain preference for bitter fruit and vegetables (cabbage, grapefruit, etc.). The distribution of these two variants changes from one population to another, but they are both present in all populations of the world. Nevertheless, their effects on food choices (notably regarding fruit and vegetables) and on health remain controversial (Feeney *et al.*, 2011; Grimm and Steinle, 2011).

These receptors are also found at other levels of the digestive tract where they modulate the response to ingested nutrients by modifying the neuroendocrine signals that regulate satiety. As a general rule, there are highly complex, individual and ethnic genetic taste patterns that have not yet been fully elucidated; they often differ between men and women but may nevertheless play a role in the stability of tastes despite contrary influences. However, tastes are largely susceptible to change according to habits, education, culture or associated environmental experiences (negative or positive).

Flavours are due to an oral somato-sensation linked to texture, temperature and irritation/pain, combined with retronasal olfaction (from the oral cavity). The perception of flavours starts during intrauterine life (the taste buds function in a liquid medium); at birth, an infant reacts positively to the flavours in his/her mother's diet, thus favouring the early transmission of familial dietary habits. These perceptions evolve quite rapidly during the different stages of childhood; they are all the more positive when complementary feeding is highly diversified. Despite some variations, these positive perceptions can be detected at the age of 18. These studies are still recent, and the research field is seeing considerable growth. Because of these marked inter-individual differences in the perception of tastes and flavours, the history of cooking (and that of the food industry) is often associated with efforts to mask certain tastes (bitterness) or enhance others (sweetness, saltiness, etc.) in order to 'mislead' natural physiological perceptions.

Sensations of pleasure are linked to palatability (viscosity, creaminess of mucilages or lipids), taste (the subtlety of sweetness), stimulation by spices or dietary variety (multiplication of visual, gustatory and palatability stimuli). This 'hedonic' response, controlled by the central nervous system, appears to encourage an increase in the quantity ingested during meals (Brondel et al., 2009; Sorensen et al., 2003).

However, it sometimes remains difficult to distinguish between innate and acquired food choices (taste, palatability, preferential appetite, etc.) in view of the cultural determinants linked to our membership for a given period to a particular social group that prefers specific foods because of their local availability and the cultural value placed upon them in this context (pleasure, rite–religion, symbol, hot–cold, etc.).

2.3.4   Animal versus plant products

Humans are naturally omnivorous; they can satisfy their nutritional needs by consuming a broad range of foods. This ability to adapt to markedly different diets according to environmental constraints, or because of cultural and social moulding, has contributed to human colonisation of the entire planet. Some extreme situations may even be observed: populations in the polar north with a strongly meat-based diet, and vegetarian individuals or groups. According to Leonard *et al.* (2010) humans are distinctive from primates and other mammals because of the size of the brain and the high proportion of energy expenditure devoted to feeding this brain. Hence the incentive to search for 'better quality' food that is dense in energy and rich in the fats and poly-unsaturated fatty acids necessary for the development and functioning of the brain; a fattier diet and an appropriately adapted digestive system (with a more developed small intestine and a shorter colon) have become our norm. There is evidence of an evolution of 'meat adaptive' genes that have enabled humans to benefit from animals as a major source of energy (Leonard *et al.*, 2010).

Different historical and ethnographic studies have suggested that the plant/animal ratio in the diet of both historical and present-day hunter-gatherers averages around 65/35 (with some broad variations); in most cases these meats are lean (except for offal) but contain higher levels of polyunsaturated fatty acids than those produced by modern livestock farmers (Mann, 2000).

Nevertheless, many populations today still live on largely plant-based diets. The American Dietetic Association and American Academy of Pediatrics agree that a well-regulated vegetarian diet can meet nutritional needs and contribute to normal growth in infants and young children. A vegetarian diet in adults generally complies with standard nutritional guidelines and will meet official recommendations with respect to different nutrients (Craig and Mangels, 2009). However, a vegetarian diet can vary in terms of its composition and may sometimes be deficient in certain vitamins (B12, D), minerals (iron, zinc, calcium) or other nutrients (certain amino acids, omega 3 fatty acids). Hence the need to monitor the growth of vegetarian children in this respect.

But in most cases, and particularly in developing countries, the consumption of vegetarian-based diets is more closely linked to local availability or financial resources, without there being any real choice or knowledge of the risks. It may therefore have a nutritional cost: iron is less bioavailable in plants, which can cause anaemia; plant phytates restrict the absorption of divalent ions (calcium, zinc, etc.), which may contribute to growth retardation in young children; proteins are naturally better equilibrated for human nutrition in animal products than in plant products (e.g. levels of lysine, which is necessary for growth) and are slightly more digestible.

For all these reasons, it is easier to satisfy nutritional needs, at least at certain ages (during growth and ageing, notably) with a minimum level of animal products in the diet (notably because they contain higher levels of vitamins and minerals), even if this is not always essential, on condition that access is ensured to a broad variety of plant products. Inversely, animal products are a more or less abundant source of lipids rich in saturated fatty acids, an excess of which is not good for health. Thus some authors have suggested a reduction in their average worldwide consumption from 100 g per day to 90 g, with a better balance between rich and poor countries (convergence) and not exceeding 50 g per day of meat from ruminants (McMichael *et al.*, 2007).

2.3.5   The role of the microbiota

The lower part of the digestive tract in humans contains a considerable biomass of micro-organisms (approximately 10^{14} bacteria) or the 'metagenome', a population which is characterised by its global genome. Despite the large number of species that may be present (between 500 and 1000), it is recognised that they are grouped into a few significant families that can affect our nutritional status, the two most important being the Bacteroidetes and the Firmicutes, which are found in different proportions in thin or obese animals or individuals. It remains to be determined whether this metagenome is universal or distributed very differently between different populations and different lifestyles, and whether it is a cause or a consequence of metabolic modifications related to obesity and its associated diseases. A certain number of studies have suggested that the flora in some subjects is more efficient in extracting energy from the diet, and this may be linked to a risk of obesity. Very recently, some authors have shown that the administration of antibiotics during the first six months of life might increase the risk of being overweight among the children of mothers with normal corpulence, but might reduce the same risk in the children of overweight or obese mothers. This effect could be explained by an impact on the establishment and diversity of microbiota. It thus appears that microbiota are an inevitable partner in the physiological effects of diet; however, numerous studies will be necessary before we understand how this diet/microbiota 'tandem' evolves and how it can be modified to promote better health (Ajslev *et al.*, 2011; Arumugam *et al.*, 2011).

Knowledge of nutrition provides a clearer understanding of how the consumption of animal products facilitates the satisfaction of physiological needs, but it does not offer a mechanism to explain or justify the very high levels of animal calorie consumption that are now the trend in many countries. Although the multiplicity of factors and mechanisms involved in regulating dietary behaviour do not enable the proposal of absolute standards, they nonetheless provide numerous opportunities to explore alternatives to the situation seen today in developed countries.

2.4   Impact of agricultural policies

Agricultural policies drive production through the support they provide to certain sectors, and in so doing they impact consumption through their action on supply. Two different streams of work have focused on the effects of agricultural policies on food consumption and their effects on health.

The first stream has been the preserve of researchers in nutrition, or more generally those working in public health. According to them, the European Common Agricultural Policy (CAP) has had a harmful effect on health by encouraging the consumption of animal fats (notably through European butter subsidy schemes) and by limiting the consumption of fruits and vegetables (notably via policies on market withdrawals). Studies have highlighted the high level of taxpayer support given to dairy production and the lack of support for fruit and vegetable production. These analyses have tended to focus on the supply side, and have notably called into question European policies that provide incentives for dairy production (Elinder, 2005; Lloyd-Williams *et al.*, 2007; Veerman *et al.*, 2006).

The second stream has been studied by economists. The approach has tended to consist in comparing the situation observed (i.e. including agricultural policies) with that which would prevail if these policies did not exist. It places considerable emphasis on the role of prices. Overall, it has led to a reverse conclusion, or at least one that is much less clear-cut. In the case of the European Union (EU), the Common Agricultural Policy (CAP) has maintained the prices of many agricultural products at levels that are significantly higher than they would be if there were no CAP, particularly with respect to dairy products and sugar. The CAP (at least before the most recent reforms) thus tended to act as a tax on fat and sugar. One study of the situation in Finland before and after it joined the EU showed that following its adhesion, the relative price of margarine had fallen in comparison to butter, and that this had contributed to a reduction in the consumption of saturated fats (Prättälä, 2003). Similarly, fruit and vegetables benefit from relatively little protection, which means that these products are available within the EU at prices that are quite similar to what they would have been without this protection. We are not aware of a global analysis in the case of the CAP, probably because support policies have varied markedly between different products. Similarly, we were not able to find any studies on the impact of recent CAP reforms. Ongoing research relative to the reform of sugar policies highlights the important role of industry in the transmission to consumers of upstream price variations. In the case of the soft drinks market, it has been shown that the food chain would benefit from a greater reduction in sugar prices, as this would cause an increase of about 5% in soft drink consumption because of reforms to sugar policies (Bonnet and Réquillart, 2010). In the case of the USA, the work by Alston, Sumner and Vosti (Alston *et al.*, 2008) suggested that the removal of support from American agriculture has had limited effects on prices, so support for agriculture cannot explain the rise in obesity (Alston *et al.*, 2006). In other words, the elimination of support for agriculture is not sufficient to guide consumers towards healthier food choices. These authors emphasised the important long-term role of agricultural research and development policies, which can have a major impact on prices. The book written by Mazzocchi, Traill and Shogren (2009) presents a review of intervention policies.

From a research point of view, it would be interesting, in the case of the EU, to gain a clearer understanding of the impact on food consumption of all the reforms that have been implemented during the past ten years and which have led (or will lead) to marked changes in the prices of agricultural products. Similarly, it would be useful to clarify the transmission of upstream price variations to the prices paid by consumers, integrating the strategic choices of food companies (in terms of price and product ranges) and distributors.

2.5   Changing trends and heterogeneity in consumption: opportunities to control future changes?

The efficiency of nutritional information is sometimes called into question, but it has some unquestionable effects. Different studies have clearly shown, for example, that scientific publications on the health impact of dietary cholesterol have heavily influenced the consumption of a certain number of foods. From 1980, there was a massive rise in the number of publications on the effects of dietary cholesterol, and this coincided with a significant trend change in the consumption of certain products of animal origin (red meat, butter, eggs, whole milk, etc.). Recent evolutions in the consumption of animal calories in European diets illustrate this change (Figure 2.7). At the end of the 1970s, all countries were seen to converge towards the level of consumption reached by the highest consuming countries (Finland and the UK), which was about 1200 kcal per person per day. But as from the 1980s, and at different rates depending on the country, a change was seen in the consumption of animal calories in northern and western European countries, together with a very marked slowing in its considerable growth in southern European countries. This convergence now seems to be leading towards a level of around 1000 kcal per person per day. Different modelling studies performed in France, Europe and the USA clearly established that nutritional information has had a significant effect, distinct from that of prices and incomes (Nichele, 2003).

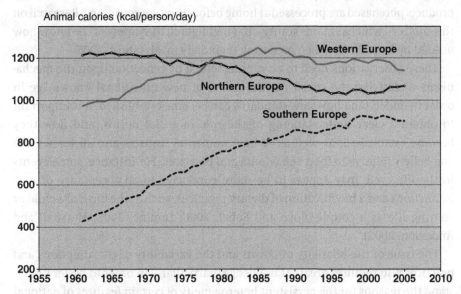

Figure 2.7 Evolution of the availability of animal calories in Europe
(Source: P. Combris, 2006. Food supply data from FAOSTAT (http://faostat.fao.org/site/609/default.aspx)).

These results showed that validated and consensual information can sustainably curb major consumption trends (Mazzocchi *et al.*, 2008; Schmidhuber and Traill, 2006).

However, work on the mechanisms governing the reception and appropriation of dietary standards has shown that knowledge of the messages is not always sufficient to achieve a change in practices. Consequent action implies the integration of practices that comply with standards in daily routines and meanings, which do not always leave space for the standards that the authorities would like to develop for public health purposes and which may even correspond to contrary norms. For example, the desire to preserve family meals in a non-confrontational atmosphere may lead some mothers in working class homes to decide not to serve vegetables as part of the evening meal (Masullo and Régnier, 2009). Furthermore, nutritional standards compete with popular beliefs concerning diet (notably with respect to the diet of children), particularly in working class households (Coveney, 2005). Research that has focused on the implementation of guidelines has thus emphasised social heterogeneity linked to differences between the constraints and the resources available to households, as much as the value systems and norms that govern them (Delormier *et al.*, 2009). A clearer understanding of social inequalities and disparities between households also requires more in-depth research on how the food products purchased are processed at home before they are consumed. Research on this subject – which remains scanty – has highlighted the variety of the know-how utilised by households when preparing their daily meals (Short, 2006).

These observations need to be supplemented by reflection upon the mechanisms underlying the social dissemination of new nutritional knowledge in order to understand how new standards can become established. It is important to obtain a clear understanding of the role of social norms, and how they become established, as well as of environmental impacts and all knowledge and beliefs relative to food behaviours and practices. For instance, some events in the life cycle may appear to be more conducive to an acceptance of new injunctions and a modification of dietary practices, such as the birth of a child or starting life as a couple (Bove and Sobal, 2006; Lamine, 2008; Marshall and Anderson, 2002).

The issue of the relativity of norms and the variability of consumptions and practices also requires greater knowledge, particularly in order to better understand the reasons for the persistent heterogeneity of certain features of national and social food models. This heterogeneity merits further analysis insofar as it may suggest sustainable alternatives to current trends.

In addition, the question of the development of knowledge, information and new social norms remains open. Consumer groups justifiably highlight the over-responsibility that may result from a purely market-driven orientation through choices and willingness to pay. Apart from the fact that such a situation introduces inequality on the basis of consumer purchasing power, it also excludes all those throughout the world who do not have access to the market. The technicality of the problems to be solved also raises the question of expertise and the risk of technocracy. Finally, the efforts made by producers to ensure greater sustainability are not sufficiently acknowledged by the market. The marketing of more sustainable products may stigmatise non-sustainable products, without increasing consumers' willingness to pay for sustainable products. If sustainable variants only benefit from a small market share, the outcome may be negative for all producers (Kanter *et al.*, 2009). All these points require combined study by economic, social and political scientists in order to determine the mechanisms and tools that are the best suited to the orientation and governance of food systems.

2.6   Heterogeneity in consumption and nutritional inequalities

The study of the convergence of food consumption models takes no account of the heterogeneity of this consumption within the same geographical region. Yet several types of consumption behaviour may co-exist within a single 'model', both in terms of the quantities consumed and the quality and diversity of the foods available (variety, choice of products from within the same range of foods). These disparities, observed in France and in other countries, have nutritional implications that are reflected particularly by social differences in health. Indeed, pathologies that are closely linked to dietary behaviour, such as obesity or diabetes, are more widespread among lower social classes (Guignon *et al.*, 2010; Mackenbach *et al.*, 2008). Furthermore, these social inequalities with respect to health are expected to increase (Charles *et al.*, 2008; Leclerc *et al.*, 2006).

The issue of nutritional inequalities is one of the ethical and operational challenges faced by any debate on food and its sustainability. Any study of major consumption trends must take account of the different equilibriums that exist according to living standards, in order to correctly understand future evolutions and for the ethical purpose of reducing inequalities. It is known that in France, the proportions of animal and plant products in the diet vary with income, the level of education or the socio-professional category, in that the majority of animal products are consumed less at the top of the social scale (Caillavet *et al.*, 2009; Recours and Hébel, 2006). Similarly, it is known that nutritional messages do not have the same impact on people with different levels of education or of different social class (Masullo and Régnier, 2009).

The convergence of diets hides very marked disparities, not only within countries but also between countries. The convergence phenomena described above are linked to economic development and are thus tributary to its fluctuations. The rise in the number of undernourished people, and the debate that has resulted from this situation, have been mentioned above. The data on which these evaluations were based were obtained from national assessments of agricultural availability, converted into calories, and they only imperfectly reflect real consumption and dietary practices. Nevertheless, they seem to indicate cases of 'backtracking' in relation to the expected progress. These observations tend to arouse reactions by different protagonists involved in food security, both private and public, national and international, in emergency situations, particularly since different crises in recent years (2005 famine in Niger; 2006–08 explosion in raw material prices) have alarmed public opinion. Finally, it should be noted that the increase in the number of undernourished people in some rich countries is starting to become a matter of debate. Over and above discussions on the concepts concerned (undernutrition, food insecurity, etc.), their measurement (food diversity, calories, perceptions, poverty, etc.) and their scale (individual, household, region, country), several questions are raised as to the resurgence of severe dietary deficiencies and imbalances, including in rich countries. The causes are poorly understood, as are their consequences, notably at the level of households and individuals: how do they adapt to a reduction in their purchasing power? Which foods are they forced to give up? What are the effects on health and well-being?

2.7   Questions for research

The different points reviewed in this chapter raise questions that are mainly linked to the difficulty of applying the diets and dietary practices of developed countries to the world as a whole. The issue of inequalities within and between countries thus lies at the heart of this debate. One of the essential points is to identify the mechanisms that could change current trends. The questions for research raised concern not only about the drivers of supply (see Chapters 4 and 5 also), for example, through agricultural policies, but also the methods of altering demand. At present, this second point seems the most complex. It implies the combination of knowledge on the individual and social determinants of behaviours, on the development of information, the construction of areas of consensus, and the role of institutions and governance regarding the process as a whole. Nine major questions thus cover the numerous issues raised in this chapter.

1. Is it possible to reduce the consumption of calories of animal origin in developed countries? Forecasts show that the current level of consumption of animal calories in developed countries is about double the level that can be applied to the whole planet. Based on the study of past trends and an analysis of current variations in consumption levels and diets, is it possible to imagine credible scenarios for a reduction? How will different food groups be affected by these changes?

2. Can we technically and legitimately set limits on the consumption of animal calories when incomes rise in emerging countries? The consumption of foods of animal origin increases as soon as the income of a population so allows. The biological, physiological, sensory and social determinants of this appetite are poorly understood. On what acceptable foundations is it then possible to base consumption limitation targets? How can technical and ethical considerations be combined to enable such an approach?

3. Which population groups will be the most affected by a decrease in animal product consumption? A reduction in the mean consumption of animal products may be accompanied by very different trends of nutritional inequalities within a population. How can the heterogeneity of changing consumption trends be characterised and their implications be measured?

4. What processes are necessary to enable the modification and social dissemination of new behaviours (reception of information and modification of social norms)? Questioning major food consumption trends implies behavioural changes that will go far beyond public health guidelines, which are already so difficult to implement. In this respect, it is necessary to carry out research that involves all the human and social sciences in order to understand how new behaviours emerge and then spread, and how they can lead to the development of new social norms. In view of the great amplitude of the changes required, consistency of information and actions will be a condition for success. This therefore implies research in order to construct a consensus on the criteria for sustainability, on the clarification of indicators and signals and finally on the harmonisation of actions. Cognitive aspects are also important: how is it possible to shift from explicit cognition (reasoned behaviour) to implicit cognition (spontaneous behaviour)? Finally, the question of social multipliers also needs to be addressed, as this is an important aspect of understanding the mechanisms for the social dissemination of behaviours and the development of new social norms.

5. Will the choice by consumers of sustainable product variants be sufficient to change food supplies? The role of the market needs to be well understood, and its potential impact must in particular be correctly evaluated. A study of the micro-decisions made by consumers, and an evaluation of their willingness to pay, will allow us to determine whether incentives for producers will be sufficient to shift supply towards greater sustainability. Similarly, the importance of consumer preferences must be assessed, particularly by better understanding the hierarchy and interactions of choice criteria. Will a simple variation in taste or price compromise the choice of a more sustainable variant?

6. What is the impact of domestic practices on the sustainability of food systems? Practically no public quantitative data are available on purchasing, storage and food preparation practices, and on the management of waste by households. The collection of such data, in both Northern and Southern countries, is therefore a priority.

7. Can policies to support agriculture and the development of international trade have an effect on the sustainability of food systems? A retrospective analysis of the effects (particularly those on prices) of different reforms relative to agriculture and the organisation of trade should be the first step in more general research on the consistency of the criteria implemented

under these policies, on the price volatility of agricultural and food products and on the impact of standards (see Chapter 8).

8. Will an acceleration of the rate of food transitions in emerging countries undermine the sustainability of food systems? Nutritionists have demonstrated the deleterious effects of the increasing rapidity of such transitions on the nutritional status of populations. The effects of this acceleration on the environment and on dietary behaviours also need to be studied. Particular attention should be paid to generational effects. Are these effects becoming increasingly marked? Do they give rise to irreversible changes in behaviours and dietary practices?

9. Whether in terms of consumption trends in different countries of the world or the factors and mechanisms that underlie the regulation of behaviours, there is considerable heterogeneity. In a context where potent forces are pushing towards convergence, it is important to study this heterogeneity and its trend. Is it possible to characterise the heterogeneity of dietary behaviours? How does this evolve at a global level and within different countries? Is this diversity destined for inevitable decline, or does it constitute a reservoir of alternatives to currently dominant trends?

Many of the questions raised above reflect the points discussed in the context of the 'Foresight' project commissioned by the British Government (Foresight, 2011; Pretty et al., 2010). The questions on limiting the consumption of animal products and changing consumption models are the most similar. The implications in terms of information for consumers, and in particular changes to social standards, focus on the same research areas. Finally, the question of the effectiveness of different systems in rewarding the efforts made by farmers (question 100 in Pretty et al., 2010) has been addressed through consumers' willingness to pay for environmental attributes.

3

Carbon footprint and nutritional quality of diets in France

Authors: NICOLE DARMON AND LOUIS-GEORGES SOLER

Contributors: ARMELLE CHAMPENOIS, CATHERINE ESNOUF, SARAH MARTIN, JÉRÔME MOUSSET, BARBARA REDLINGSHÖFER, MARIE RUSSEL, MARKÉTA SUPKOVA, DJILALI TOUAZI AND FLORENT VIEUX

This chapter presents some original findings based on a study of the carbon footprint of food in France. Factors that are liable to explain the inter-individual variability of the carbon footprint associated with the usual food consumption of a representative sample of French adults were explored. By analysing the relationship between the nutritional quality of food and its carbon footprint, this chapter also addresses the more general issue of the compatibility of two of the pillars of sustainability.

The authors would like to thank the ADEME and INRA for their financial support that enabled this study to be conducted.

3.1 Introduction

Initial observations of the environmental impact of food demonstrated a broad inter-individual variability of this impact (even when taking account of variations in energy consumption). This contributed to the suggestion that dietary choices could simply be altered (i.e. replacing certain foods) to reduce the environmental impact of food (Carlsson-Kanyama et al., 2003; Coley et al., 1998). In particular, a reduction in the consumption of red meat from ruminants was suggested, because it is the production of these foods that generates the most greenhouse gas emissions per kilocalorie (kcal) (approximately 11 g eqCO$_2$/kcal) (Kling and Hough, 2010). It is true that livestock farming accounts for 80%

Food System Sustainability: Insights from duALIne, eds. Catherine Esnouf, Marie Russel and Nicolas Bricas. Published by Cambridge University Press. © Cambridge University Press 2013.

51

of greenhouse gas (GHG) emissions in the agricultural industry (Steinberg *et al.*, 2006). For this reason, plant-based foods and vegetarian diets have been shown to exert a lower environmental impact than omnivorous diets (Baroni *et al.*, 2006; Carlsson-Kanyama and Gonzalez, 2009; Marlow *et al.*, 2009; Reijnders and Soret, 2003; Risku-Norja *et al.*, 2008).

Furthermore, the beneficial effects on health of a diet rich in fruits, vegetables and other complex plant products, and only containing moderate quantities of animal products, have been demonstrated on many occasions, while a diet qualified as 'Westernised' – characterised not only by a high consumption of red meats and processed meats, but by an excessive consumption of refined cereals, fried foods and sweet products – is associated with higher rates of total mortality and chronic diseases (WHO, 2003). Because first, plant products have a lower environmental impact than animal products, and second because a diet rich in plant products appears to provide protection against chronic diseases, it is now almost universally accepted that a global change towards a diet mainly made up of plant products would have a favourable impact on both the environment and health (Duchin, 2005; Garnett, 2011; Tukker *et al.*, 2011).

More specifically, international convergence in the meat consumption (reduction in high-income countries and a reasonable increase in low-income countries) has been proposed as a good way to reduce GHG emissions and to improve the health of populations (Griffon, 2006; McMichael *et al.*, 2007; Paillard *et al.*, 2010).

However, it is possible that the good health of vegetarians may owe more to their interest in their own health, which has led them to adopt behaviours more in line with recommendations on physical activity, a well-balanced diet and smoking, than to the fact that they do not eat any meat (Key *et al.*, 2006). In addition, meat and fish, and to a lesser extent dairy products, are irreplaceable sources of some essential nutrients, and reducing their consumption constitutes a real nutritional challenge (Millward and Garnett, 2010). In fact, a diet that includes very small quantities of animal products can be healthy if, and only if, it is otherwise perfectly balanced. It all therefore depends on the substitutes that are used to limit environmental impacts (Vieux *et al.*, 2012). In particular, it appears that foods rich in fats or sugars have a low carbon impact (approximately 2.5 g eqCO$_2$/kcal) (Kling and Hough, 2010). The relatively low carbon impact of unhealthy foods has recently been confirmed by French data, which suggest the existence of possible incompatibilities between environmental and nutritional objectives (Vieux *et al.*, 2013). Similarly, numerous publications have underlined the need to increase the proportion of fish in the diet to attain the levels determined by nutritional guidelines (Arnoult *et al.*, 2010; Maillot *et al.*, 2010), but the sustainability of these high consumption levels is much criticised from environmental, economic and toxicological points of view (Lang, 2005; Lobstein, 2002).

Consequently, the recent FAO report recommended that sustainability should be taken into account when compiling dietary guidelines and nutritional policies, and emphasised the need for studies that would demonstrate synergies between the different dimensions of sustainability (FAO, 2010b). It is therefore both essential and urgent to study the relationship between the nutritional quality of food and its environmental impact, in order to identify the choices that will enable a reduction in environmental impact while taking account of other aspects of a sustainable diet, and in particular its cultural and economic acceptability and nutritional adequacy.

This study had two main objectives:

- first, to estimate the carbon footprint of the diets usually consumed in France
- and second, to analyse the carbon footprint of diets *according to its nutritional quality*.

By comparison with other international studies which have analysed average or stereotyped consumption patterns (meat-based diets, vegetarian diets, for example), the approach adopted here is doubly original because:

- it considers the spontaneous food consumption of individual members of the general population
- it defines nutritional quality as compliance with nutrient-based recommendations rather than on a food-based approach of the composition of a balanced diet.

3.2   Quantification of the carbon footprint of diets

3.2.1   Methods

The preliminary step in this analysis consisted of the collection of consumption data. In any study, this step is crucial as it influences data quality. Several methods are available at present. No reference method really exists, and each method has its advantages and disadvantages. The choice of a collection method needs to be compliant with the type of study and its specific objectives (Etievant *et al.*, 2010; Zetlaoui *et al.*, 2011).

The method used to group foods into categories (i.e. nomenclatures) is also important, particularly when combining variables belonging to different areas of study and different disciplines. This is the case for the combination of nutrition, the environment and economics. For surveys on individual food consumption and nutritional epidemiology, foods are grouped according to their nutrient composition. In purchasing surveys, foods are grouped according to how they are organised in retail outlets. For work on environmental impacts, it would probably be wise to classify them by production type, packaging, storage, etc. In order to analyse the relationships between these different dimensions, a common nomenclature would need to be created, which would either be very rough (lowest common denominator) or would have to be very detailed (all information available on each food in terms of its production, sale, consumption, etc.).

For the purposes of the present analysis, we adopted a 'nutritional' nomenclature because our work was based on the INCA 2 study, whose initial objective was to describe individual consumption patterns and analyse the nutritional intakes of the French population.

3.2.1.1   Identification of 'representative' foods

This study was based on data on the food consumption of representative adults (1918 registered individuals aged 18 years and over: 776 men and 1142 women) who participated in the INCA 2 survey (Individual National Study of Food Consumption in France) performed in 2006–07 by the French Food Safety Agency (then AFSSA, now ANSES) (Lafay and Volatier, 2009).

A total of 1312 different foods were declared as having been consumed during the week of the survey by all the participants in the INCA 2 survey. Data concerning the carbon footprint of the foods were rare and difficult to obtain (generally privately held), so it was impossible to consider collecting data on the carbon footprint of each of these 1312 foods within the timeframe of the study. The present analysis is thus based on a more limited number of 'representative' foods that are widely consumed by the French population. Foods were deemed to be 'representative' if they were mentioned by the largest number of consumers compared to products with similar nutritional characteristics (i.e. belonging to the same family (n = 36 families) and the same category (n = 11 categories)). As far as possible, we selected foods that were consumed at least once during the week of the survey by at least 10% of the adults. The real or supposed availability of environmental data on these products was also one of the choice criteria, as were the specific nutritional qualities of certain foods (e.g. nuts, lentils, sardines).

3.2.1.2   Carbon footprint of each 'representative' food

The term 'carbon footprint' designates the global warming potential of emissions of the six greenhouse gases (GHG) listed by the IPCC. The carbon footprint of a diet is calculated from the carbon footprint of different foods, which in turn is estimated from life cycle assessments (LCA) of the products involved. Life cycle assessment is a standardised method (ISO 14044) used to quantify the environmental impacts generated by a product throughout its life cycle. It provides an estimation of the 'carbon footprint' of foods available for sale, which, in equivalent CO_2 (eqCO_2), designates the quantity of GHG emitted by producing, processing and transporting the products. The main gases are carbon dioxide (CO_2) which results from the combustion of fossil fuels, methane (CH_4) produced by enteric fermentation in ruminants and animal waste, and nitrous oxide (N_2O) associated with nitrate fertilisation and the management of animal waste. Greenhouse gas emissions (or the carbon footprint) compiled for the different food products covered by the study were expressed in g eqCO_2/ 100 g of product.

Data on the carbon footprint of each representative food were thus determined, taking account of the following stages: agricultural production, processing, packaging, transport to a retail outlet in France and sale, including storage by the distributor. The stages involved in transport between the point of sale and the consumer's home, consumption and use during storage and preparation by the consumer, as well as management of the end of life of the food product, were not taken into account due to a lack of easily available and reliable data. The functional unit chosen for the collection of carbon data was 100 g of the food product 'as purchased' (i.e. available in a retail outlet) in France.

The carbon footprints of the representative foods were either determined from the carbon footprints available in the international literature, French and international LCA studies, theses, scientific articles and industrial data, or resulted from a reconstitution. For most of the products analysed, a mean carbon footprint was calculated in order to define the final carbon impact. Extreme carbon footprints were deliberately discarded.

Once all the carbon footprints collated in this way had been validated by the Monitoring Committee, the carbon footprints of the products 'as purchased' had to be converted into carbon footprints of products 'as consumed'. For example, the impact of meat purchased on the bone was converted into the impact of meat off the bone by applying a conversion factor (the edible portion) taken from the *CIQUAL French Food Composition Table* (1995). Given the diversity of production modes, distribution channels and packaging types, the scope of the analysis had to be restricted choosing as 'representative' foods only those from conventional agriculture and that were produced in France, except for tropical fruits such as bananas or oranges, imported and then distributed by lorry to retail outlets throughout France.

3.2.1.3 Daily carbon footprint of individual diets

Based on individual consumption data, carbon footprint values for 73 representative foods, and hypotheses regarding the proportions of these 73

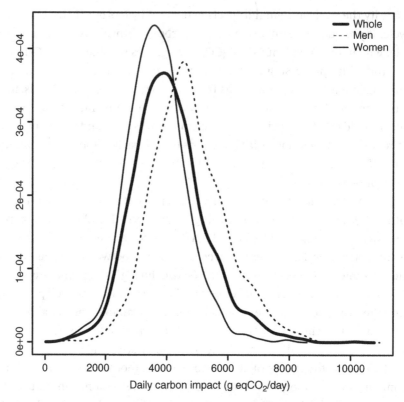

Figure 3.1 Distribution of the daily carbon footprint (g eqCO$_2$/day) of the diet of adult participants in the INCA 2 study (total population: men and women).

foods in individual consumption, it was possible to calculate the daily carbon footprint of the diets of each of the 1918 individuals involved in the study.

3.2.2   Results

3.2.2.1   Carbon footprint of the diet: mean and inter-individual variability

Figure 3.1 shows the distribution of daily carbon footprints within the INCA 2 population sample. Considerable inter-individual variability can be seen around these means. The mean carbon footprint of the diet of adults in France was 4090 g eqCO$_2$ per person per day (standard deviation: 1175), with 4725 g eqCO$_2$/day for men (standard deviation: 1183) and 3658 g eqCO$_2$/day for women (standard deviation: 953).

The finding that the diet of men had a greater carbon footprint than that of women may have been due to the fact that men eat more than women, or because they eat differently (in particular, men eat more meat and fewer fruits

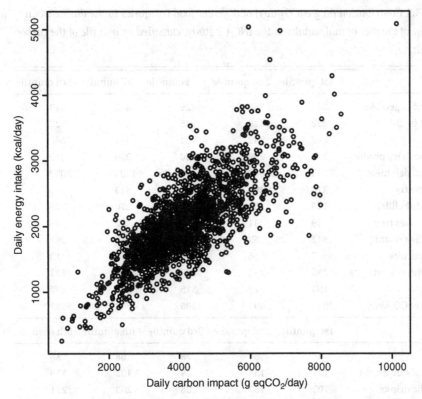

Figure 3.2 Relationship between daily caloric intake (kcal/day) and the daily carbon footprint of the diet (g eqCO$_2$/day) of adults participating in the INCA 2 study.

and vegetables than women). More generally, the marked inter-individual variability observed regarding the carbon footprints of the diets may have been due to either a 'quantity' effect or 'structural' effect.

Figure 3.2 reveals a very strong positive correlation between the total calories ingested by individuals and the daily carbon footprint of the diet. Similarly (data not shown), a very strong positive relationship could be seen between the total quantities consumed and the daily carbon footprint.

3.2.2.2   Contribution of each food category to the diet's carbon footprint

The participants were grouped into quintiles, taking men and women separately, according to the daily carbon footprint of their diet. The contribution of each food category to the daily carbon footprint was then calculated by quintile. Tables 3.1 and 3.2 show how, for men, the contributions of each food category evolved in terms of absolute values (g/day) and as a percentage, depending on the quintile of the total carbon footprint:

Table 3.1 *Contribution (in g eqCO$_2$/day) of different food categories to the daily carbon footprint of the diet of male adults in the INCA 2 study, classified by quintile of the carbon footprint.*

MEN	1st quintile	2nd quintile	3rd quintile	4th quintile	5th quintile
Fruit and vegetables	312	369	429	442	513
Starchy foods	222	223	245	276	320
Cheese	247	342	421	481	642
Milk and dairy products	217	230	240	270	317
Meat and deli meat	743	1008	1256	1534	2032
Eggs, poultry	210	293	295	313	366
Fish and shellfish	124	152	191	270	328
Mixed dishes (veg.)	39	41	35	32	46
Mixed dishes (ani.)	342	362	373	375	397
Salted biscuits	7	6	8	11	13
Sweet and desserts	286	384	399	410	511
Fat	193	279	315	389	497
Total (g eqCO$_2$/day)	2942	3691	4206	4804	5981

MEN	1st quintile	2nd quintile	3rd quintile	4th quintile	5th quintile
Water	46	64	78	68	88
Non-alcoholic drinks	121	138	129	132	155
Alcoholic drinks	102	167	208	263	271
Total (g eqCO$_2$/day)	269	369	415	463	515

- In the whole population (men and women together), the daily carbon footprint of the total diet (including water and other drinks) ranged from 2900 g eqCO$_2$/day for the first quintile to up to 6063 g eqCO$_2$/day for the fifth quintile.
- The contribution in absolute values of each food category to the total carbon footprint increased in line with the total carbon footprint. For example, red meats and processed meats contributed between 743 and 2032 g eqCO$_2$/day, sweet products and desserts between 286 and 511 g eqCO$_2$/day, and fruits and vegetables between 312 and 513 g eqCO$_2$/day, from the first to the fifth quintile of the carbon footprint, respectively.
- The proportional contribution of different food categories was not totally homogeneous from the first to the fifth quintile. A reduction in the contribution of prepared foods containing ingredients of animal origin (12% to 7%), staple foods (8% to 5%) and dairy products (7% to 5%) was recorded, as well as a rise in the contribution of red meats and processed meats (25% to 34%), cheese (8% to 11%) and alcoholic drinks, when the total carbon footprint increased.

Table 3.2 *Percentage contribution of different food categories to the daily carbon footprint of the diet of male adults in the INCA 2 study, classified by quintile of the carbon footprint.*

MEN	1st quintile	2nd quintile	3rd quintile	4th quintile	5th quintile
Total (g eqCO$_2$/day)	2942	3691	4206	4804	5981
Fruit and vegetables	11	10	10	9	9
Starchy foods	8	6	6	6	5
Cheese	8	9	10	10	11
Milk and dairy products	7	6	6	6	5
Meat and deli meat	25	27	30	32	34
Eggs, poultry	7	8	7	7	6
Fish and shellfish	4	4	5	6	5
Mixed dishes (veg.)	1	1	1	1	1
Mixed dishes (ani.)	12	10	9	8	7
Salted biscuits	0	0	0	0	0
Sweet and desserts	10	10	9	9	9
Fat	7	8	7	8	8
Total (%)	100	100	100	100	100

MEN	1st quintile	2nd quintile	3rd quintile	4th quintile	5th quintile
Total (g eqCO$_2$/day)	269	369	415	463	515
Water	17	17	19	15	17
Non-alcoholic drinks	45	37	31	29	30
Alcoholic drinks	38	45	50	57	53
Total (%)	100	100	100	100	100

The results shown in Tables 3.1 and 3.2 only concern men, but similar trends were observed among women.

In all the carbon footprint quintiles, the food category that contributed the most to the diet's carbon footprint, in both absolute and relative terms, was that of red meats and processed meats, thus confirming the results of most international studies.

3.2.2.3 Variability of the diet's carbon footprint: 'structural' effect or 'quantity' effect?

It was necessary to determine whether the relationship observed in Figure 3.2 (correlation between kcal and the carbon footprint) reflected a 'quantity' effect linked to the total volumes consumed (individuals with the highest carbon footprint consumed more of everything, as shown in Table 3.1) or a 'structural' effect (where differences between the proportions of foods, as seen in Table 3.2, explained the variability in individual emissions, more clearly than the total quantities ingested did). To clarify this issue, the daily carbon footprint (CF) of the diet of each individual (in g eqCO$_2$) can be broken down as follows:

$$CF = EF * ED * Q$$

where EF represents the carbon footprint of the diet expressed as calories ingested (g eqCO$_2$/kcal), ED is the energy density of the diet (kcal/kg) and Q is the total quantity consumed by the individual (kg/day). The first two terms reflect the 'structure' of the diet (linked in part to each category of foods, as a percentage of individual daily consumption). By definition, the third term refers to the quantity (which may cause a variation in individual CF values, when the structure of food consumption is fixed). The question that interested us here was the extent to which a variation in individual CF values within the population could be explained by structural effects (the percentage of each food category in the daily diet) or quantity effects (the total weight of the daily diet).

To answer this question, individual CF values were regressed on each of these three variables. As a result, *the effects were ranked by decreasing importance as follows: 'quantity' effect, then the 'carbon footprint by calories ingested', then the 'energy density' effect* (Vieux et al., 2012).

3.3   The choices: carbon footprint and nutritional quality of food

3.3.1   Definitions of a balanced diet

A balanced diet is a concept that is fully accepted by the general public. Scientists refer to it frequently and it is widely used by both healthcare professionals and different stakeholders in the food sector. *However, there is no consensus as to the definition of a balanced diet.* The food-based dietary guidelines that have developed throughout the world to promote the healthiest food choices, such as those issued by the EFSA and WHO (EFSA, 2010; WHO, 1998), all differ from each other. The definition of recommended food groups, as that of recommended quantities, varies from one country to another. Even advice on the consumption of fruit and vegetables may vary between countries and expert committees issuing it, in terms of both quantity and diversity. As there is no consensus regarding the definition of a balanced diet, *there is no single way to estimate overall diet quality* (Kant, 1996; Waijers et al., 2007). In practice, numerous indicators co-exist and are widely used in nutritional epidemiology to demonstrate relationships between food and health (Willet, 1998). These indicators are generally based on official recommendations concerning food and/or nutrient intakes, but there is no recognised 'gold standard' and the objectives pursued may differ markedly.

Even the highly publicised 'Mediterranean diet pyramid' (Willett *et al.*, 1995) has been the subject of differing interpretations, as attested by the existence of different scores for compliance with a Mediterranean-style diet. Thus the consumption of meat and dairy products is deemed to be unfavourable[1] to overall diet quality by the Mediterranean Diet Score (Trichopoulou *et al.*, 2003), while other 'Mediterranean' scores consider moderate consumption of these foods to be favourable[2] (Goulet *et al.*, 2003), or do not take account of them at all (Issa *et al.*, 2011).

This multiplicity of food guidelines and quality scores suggests that *there is no single good diet but no doubt very many of them*. Studies based on the modelling of individual diets have indeed shown that nutritional needs can be met in several ways by combining a wide selection of foods (Maillot *et al.*, 2010).

In order to comply with nutritional guidelines, a minimum calorie intake is required (Darmon and Briend, 2001), as is a minimal budget (Darmon *et al.*, 2006) and the presence of some irreplaceable foods (fruits, vegetables, pulses, whole cereals, nuts, oily fish) (Maillot *et al.*, 2009). Once these requirements have been met, it can be said that there are as many ways of meeting nutritional needs as there are ways of eating.

Thus, for a given individual, the correct diet is one that enables him (or her) to meet all his nutritional needs while complying with his food preferences and budget, as well as his social and cultural environment. In a given region, the dietary balance that should be encouraged is the one that takes account of availability and local customs. For this reason, a team of Norwegian researchers recently questioned the universal promotion of the Mediterranean diet, insisting on the contrary on the importance of taking account of regional specificities in the development of dietary guidelines (Bere and Brug, 2009).

Although they may also vary from one country to another, *nutrient intake recommendations show greater consensus than recommendations on food intake*. For this reason, in the study described below, we decided to base our work on the level of compliance of nutritional intakes with guidelines in order to identify balanced diets.

The advantage of this approach is that it can be transposed from one region to another, while the application of national guidelines or the 'Mediterranean

---

[1] Allocation of negative points when the consumption of dairy products and meat is higher than the respective medians in the population studied.
[2] Allocation of negative points when the consumption of dairy products and meat is too low or too high.

pyramid' cannot. Thus, based on nutritional guidelines rather than on presuppositions concerning a 'balanced diet', it is possible to identify diets that are adequate from a nutritional point of view and respect the extraordinary individual, economic, geographical, social and cultural diversity of food consumption patterns.

3.3.2   Methods

3.3.2.1   Choice of nutritional quality indicators

To estimate the nutritional quality of a diet, nutritionists use the MAR (Mean Adequacy Ratio), which estimates the mean percentage adequacy of a certain number of nutrients relative to the recommended intakes of these nutrients (Madden et al., 1976). This percentage is negatively correlated to the energy density (ED, in kcal/100 g) of the diet, the latter being considered as an indicator of poor nutritional quality (Ledikwe et al., 2006). The availability of synthetic indicators of nutritional quality, whether positive or negative, is of considerable value as they can establish correlations with other dimensions of the diet. Thus the MAR and ED have demonstrated a positive association between the nutritional quality of food and its cost (Maillot et al., 2007). By analogy with the MAR, in the present study we also defined the MER (Mean Excess Ratio), which estimates the excess nutrients (compared with the maximum recommended values), whose intakes should be limited in a balanced diet (sodium, saturated fatty acids and added sugars).

3.3.2.2   Identification of food consumer groups differentiated according to the nutritional quality of their diet

The MAR, MER and ED were calculated for each individual diet. The individuals were then classified based on the values for each indicator in relation to their observed medians in populations of men and women taken separately.

In this way, four groups of food consumers were identified based on the nutritional quality of their diets (cf. Figure 3.3):

- High nutritional quality diets, which lie at the intersection of three criteria of good nutritional quality: a MAR higher than the median, a MER lower than the median and an ED lower than the median. This included 101 men and 180 women.
- Low nutritional quality diets, which do not meet any of these criteria (108 men and 178 women).
- Intermediate+ diets, complying with two out of three criteria for good nutritional quality (294 men and 389 women).
- Intermediate– diets, complying with only one criterion of good nutritional quality (273 men and 395 women).

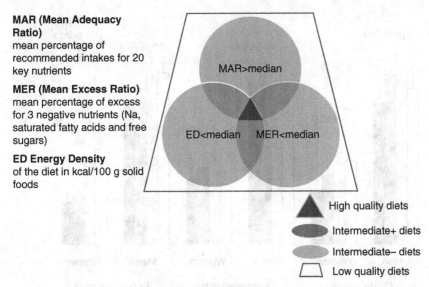

MAR (Mean Adequacy Ratio)
mean percentage of recommended intakes for 20 key nutrients

MER (Mean Excess Ratio)
mean percentage of excess for 3 negative nutrients (Na, saturated fatty acids and free sugars)

ED Energy Density
of the diet in kcal/100 g solid foods

MAR>median

ED<median   MER<median

△ High quality diets

◗ Intermediate+ diets

⬭ Intermediate– diets

▱ Low quality diets

Figure 3.3 Exclusively nutritional indicators (not based on stereotyped diet compositions) used to qualify the nutritional quality of diets.

For each gender, there were thus four groups of food consumers, and for each group we were able to calculate not only consumers' food and nutrient intakes but also their carbon footprint.

This gave us eight groups of individuals differentiated according to the nutritional quality of their diet: four for men and four for women.

3.3.3   Results: carbon footprint of the diet according to its nutritional quality

Figure 3.4 shows that the *carbon footprint was little influenced by the nutritional quality of the diet.* Among men, the carbon footprints of the four nutritional quality groups did not differ (p = 0.1596). By contrast, among women, the difference was significant, whereby *low nutritional quality diets had a smaller carbon footprint* than high nutritional quality diets. The results therefore did not follow the expected trend of healthier diets having a smaller environmental impact.

Adjustment of these raw results for energy intakes or the quantities consumed showed that:

- when the total quantity consumed was 'fixed', the groups with high nutritional quality diets (men and women) had the smallest carbon footprint;
- when the energy intakes were fixed, the groups with low nutritional quality diets (men and women) had the smallest carbon footprint.

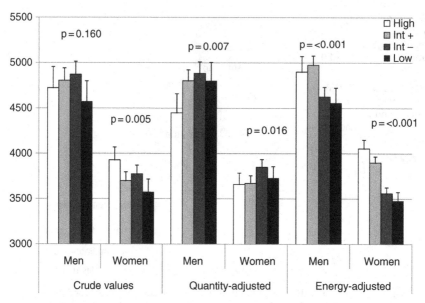

Figure 3.4 Mean carbon footprint (g eqCO$_2$/day) by nutritional quality group and gender, with or without adjustment by quantity and energy intake. The analyses were performed separately for men and women.

This is due to the fact that consumers of high nutritional quality diets ingest greater quantities of foods than consumers of low nutritional quality diets, as shown by the INRA-ADEME report in 2010 (Supkova *et al.*, 2010), because nutritional quality is inversely correlated to the energy density of diets. Thus, even although high nutritional quality diets were characterised by a predominance of foods with a smaller carbon footprint, such as staple foods, fruit and vegetables, the fact that they were eaten in larger quantities explains why, ultimately, these diets had as much, if not more, impact than low nutritional quality diets.

3.4   Limitations of the study and outlook

3.4.1   *Limitations*

The INCA 2 table, comprising 1312 foods, was shrunk down to a table of 73 'representative' foods grouped into 36 families. In fact, these foods only represented half of the quantities consumed and a third of the energy ingested; for this reason, hypotheses regarding the representativeness of these foods in the diet had to be established. A test was performed to try and estimate the error linked to this contraction. This test consisted in reclassifying individuals, according to the nutritional quality of their diets, into high quality, intermediate+, intermediate– and low quality diets, by calculating the MAR, MER and ED based not on all the foods consumed (i.e. 1312 foods, as was done to identify the eight groups of consumers) but on the 73 representative foods only. The results showed that despite this simplification, 67% of individuals remained within the same nutritional group of food consumers and 96% in the same group or an immediately neighbouring group. In addition, similar conclusions were drawn by a recent French study based on large quantities of food carbon data (Vieux *et al.*, 2013).

Life cycle assessment (LCA) is an approach that is widely used to analyse the environmental impact of a product or process. Its principal limitation is the potential influence that the initial hypotheses and methodological choices made throughout the study (scope, functional unit, impact categories, allocation rules, etc.) can have on the final results. The approaches adopted for the LCA of the food products selected in this study may differ between each other and relative to the scope of the study concerned. In particular, different food sectors can be markedly affected by the issue of allocations (of by-products), which often concern upstream, agricultural stages and initial processing, which also give rise to hypotheses specific to each study. The calculation methods employed may also vary, thus preventing comparisons with results from the different studies available in the literature.

Independently of methodological choices, GHG emissions by agriculture result from biological processes and thus vary considerably according to the environment (geographical and temporal variability) and production methods. Furthermore, the data available rarely included variations in carbon stocks in the soil (grasslands, etc.) and in plants themselves (fruit trees, grapevines, etc.), which may considerably modify any estimates of GHG emissions.

The GHG data available thus involve a *strong degree of uncertainty*, which requires considerable caution when interpreting results. A more precise analysis of relationships between the environment and the healthiness of foods will be possible through the development of public LCA databases on agricultural and then processed products.

3.4.2   Outlook

In order to generate more reliable results and hence more robust conclusions, greater accuracy would be required in estimating the carbon footprints of product families and diets. This improved precision could be achieved first, by increasing the number of bibliographical references relative to the carbon impacts of foods, and second by increasing the number of representative foods (in which case, new foods should be chosen preferentially from the families with the greatest contribution to energy intakes and to diet-related carbon footprints).

It would be interesting to compare the differences between the impact of actual current consumption and that of a high-meat diet, a vegetarian diet or a *diet complying with the French food-based dietary guidelines (PNNS) or French recommendations on nutrient intakes (ANC)* in order to evaluate the compliance or contradiction between guidelines and the carbon footprint. The challenge is to identify the driver that can be effectively triggered to enable a shift towards greater sustainability of food supplies; i.e. consumption or other links in the chain.

Given the small number of LCA studies available, it is currently difficult to determine the carbon footprint of the storage and preparation/cooking stages of each product. However, a generic value could be defined for each cooking method associated with a product group. Taking account of this impact in the combination of nutritional data and the quantities ingested (and not the quantities of raw products) would appear to be pertinent. The carbon footprint of prepared dishes upon purchase already takes account of their culinary preparation (industrial or small-scale). To ensure consistency, the carbon footprint of the home cooking/ preparation of raw products should be integrated in the carbon footprint of each product requiring culinary preparation (e.g. beef burgers, pasta, rice, etc.) before they are consumed.

As for travel between retail outlets and consumers' homes, this stage was not taken into account during our study. Its carbon footprint varies, depending on the type of transport and the distance travelled by the consumer. Hypotheses for different transport types could be advanced, based for example on the ADEME *Guide des facteurs d'émission* (Guide to Emission Factors, ADEME, 2010), in order to establish their related carbon footprints and supplement the scope of the study.

The impacts of the transport and packaging type of each product could also be studied in more detail. Through such analysis, the impact of products that are imported out of season could be studied in relation to their storage, which requires energy and refrigerant gases that contribute to global warming.

Subsequently, environmental challenges other than the carbon footprint (biodiversity, water, soil, etc.) will need to be considered.

This study showed that inter-individual variations in the carbon footprint of different diets depended on the total quantities consumed and the consumption structure.

If these findings were to be confirmed by those of other studies, this would mean that a reduction in the carbon footprints of diets may be envisaged by both lowering the total quantities consumed and reducing the carbon footprint associated with each calorie ingested. This latter objective could indeed be achieved by either reducing the environmental impacts of each family of foods through actions on the supply side, or by replacing foods with a high 'carbon footprint/calorie' ratio with foods that display a lower ratio.

3.5   Conclusion

The view that plant products are good for health and the environment, while animal products are bad for both these dimensions, seems simplistic and needs to be reconsidered. In this respect, particular attention should be paid to the impact – on the results of studies and on their dissemination – of how foods are categorised in the analyses. For red meats and processed meats, notably, very different conclusions can be drawn according to whether a group is made up of delicatessen products, hamburgers, sausages and other processed meats (Gonzalez and Riboli, 2010; Micha *et al.*, 2010).

With respect to LCA and other methods used to analyse impacts, the study highlighted the same infrastructure requirements as those described later in Chapter 10 (Methods). Apart from an obvious need for data that are both horizontal (product range) and vertical (at all levels of the chain), this study highlighted the lack of consideration for transport, packaging, domestic practices and of a large number of indicators; it stresses the need to take these aspects into account, at both conceptual and methodological levels.

The crucial issue is to determine the extent to which changes to consumption behaviour could enable a significant reduction in GHG emissions, and particularly the conditions under which such changes might be acceptable and ultimately implemented. Without going too far on this point, it is nonetheless possible to identify or situate some potential drivers. To what degree will improvements in environmental performance on the supply side, from the production of raw materials to the finished product, enable a reduction in GHG emissions, while at the same time reducing the need to modify eating patterns? This is an important question for debate and research in the future concerning diet and the sustainability of food systems.

The results of this study direct questions to public decision-making and the orientation of research at the interface between nutrition and the environment: how far are nutritional guidelines and environmental demands compatible? Indeed, the study showed that the relationship between the nutritional quality of individual diets and their carbon footprints was weak, partly because high nutritional quality diets contain large quantities of foods with a small carbon footprint. It was also seen that in order to reduce the carbon footprint of diets, one important option would be to reduce the consumption of calories, independently of the food families concerned. To what extent are consumers ready to make such changes, under which conditions and within what timeframe? It can be noted in this respect that to fight obesity and its associated diseases, all official bodies, starting with WHO (2003), insist upon the need to encourage an energy balance, first through regular physical exercise and second by consuming a low energy density diet to limit the risks of consuming more energy than required. In addition, frugality is a recommendation that forms an integral part of many food guides, starting with the Mediterranean pyramid (Willett *et al.*, 1995). Furthermore, a link between health status and compliance with the

French PNNS guidelines was only found when a penalty was attributed to individuals with excessively high-energy intakes (Estaquio *et al.*, 2008). Finally, in the USA, a report issued by several learned societies proposed the adoption of a gradual, 'small steps' approach in order to reduce the obesity epidemic over the long term. Among the recommendations made in this context, it was suggested that each person should walk 2000 steps more each day, reduce the energy density of the diet by 10 kcal per 100 g of food consumed and, above all, reduce energy intakes by 100 kcal/day (Hill, 2009).

4

Food systems

Authors: PAUL COLONNA, STÉPHANE FOURNIER AND
JEAN-MARC TOUZARD

Contributors: JOËL ABECASSIS, CÉCILE BROUTIN,
DIDIER CHABROL, ARMELLE CHAMPENOIS,
CHRISTIAN DEVERRE, MARTINE FRANÇOIS, DANIELLE
LO STIMOLO, VANESSA MÉRY, PAULE MOUSTIER AND
GILLES TRYSTRAM

This chapter focuses on food systems using an innovative dual approach. First of all, by considering them in interaction with energy and chemical systems within ecosystems, it poses the question of resource allocation (land and biomass). Second, a socioeconomic approach then highlights the diversity of these food systems. Different types of systems co-exist and reflect different ways of producing, processing, distributing and consuming food products. The 'global' food system is a constantly changing combination of these different types of systems, all of which influence each other.

Thanks to this dual approach to food systems, new research questions have emerged. New analytical frameworks would enable a clearer understanding of the interconnections between food systems within ecosystems, on the one hand, and their diversity and constant recombination on the other.

4.1   Introduction

Referring to the pioneering work of Malassis (1996), Rastoin *et al.* (2010) defined a food system (FS) as

> *an interdependent network of stakeholders (companies, financial institutions, public and private organisations) localised in a given geographical area*

Food System Sustainability: Insights from duALIne, eds. Catherine Esnouf, Marie Russel and Nicolas Bricas. Published by Cambridge University Press. © Cambridge University Press 2013.

(region, state, multinational region), participating directly or indirectly in the creation of a flow of goods and services geared towards satisfying the food needs of one or more groups of consumers, both locally and outside the area considered.

The discussions initiated in the context of the duALIne project considered food systems in terms of both supply and demand relative to the three pillars of sustainability. This definition is in fact very close to the accepted meaning of 'food systems' proposed in the English-language literature to represent 'all processes involved in feeding a population, and encompassing the input required and output generated at each step. A food system operates within, and is influenced by, the social, political, economic and environmental context' (Goodman, 1997). This notion may be associated with political values that are reflected by classifications such as 'conventional food systems' (industrial) and 'alternative food systems' (e.g. local food systems, organic food systems, fair trade, etc.) proposed as opportunities for sustainable development. By including agricultural production and food processing, this definition is distinctive from proposals that tend to limit the food systems to the 'final' activities of supply and consumption by individuals and households (Branger *et al.*, 2007).

Over and above these differences, we emphasise that the technological stages of any food system remain qualitatively the same under the different approaches: storage, fractionation, functionalisation, formulation, final application – with functionalisation being perhaps considered as the only new technological contribution of the twentieth century in terms of either ingredients or additives. Only the scale of the equipment involved, and the distribution and number of operators between a unit of place and time (at the scale of a family), and a multiplication of operators at numerous sites (at the industrial scale) will allow a distinction between several systems.

The definition of a food system is transversal and more general than several notions describing the practical activities and flows affecting a specific agricultural or food product: an agricultural or food chain which formally identifies all operations, flows and actions involved from the agricultural production to consumers; the supply chain that lays emphasis on the organisation of these operations and flows from the standpoint of supplying an operator (or group of operators); the value chain that covers the sequence of activities that create value, considered from the perspective of a company (Porter, 1996); the distribution network or channel focused on the exchanges or services associated with a given product (without considering its major transformations upstream or downstream), etc. Of these notions, that of the chain is easy to comprehend as it is based on the additivity of technical actors throughout the flow of molecules, from production to processing. This approach is useful in that it reflects the characteristics of agricultural raw materials, which are biologically and chemically unstable, unsafe, have variable qualitative characteristics and are of fluctuating nutritional quality. The development of new agro-ecological practices, the introduction of new varieties resistant to biotic and abiotic stresses, the diversification of cultivated plants and livestock and new uses for agricultural by-products will lead to *greater variability of the raw materials available*. Food technologies transform these raw materials into food products that are controlled, more stable and of good health quality. Industrial and small-scale production will become increasingly differentiated relative to their approaches. While small-scale farmers and processors depend on their know-how and adaptive ability, industrial companies will systematise measurements to reduce the subjectivity of interventions and guarantee the qualitative stability of foods. By contrast, the 'chain' notion poorly reflects a major characteristic of foods: their structural complexity resulting from the managed interactions of different ingredients and additives obtained from different chains.

Broader than the notion of the 'chain', that of a 'food system' satisfactorily reflects all the resources, institutions, practices and stakeholders through which societies organise their food supplies (Rastoin *et al.*, 2010).

This chapter aims to analyse factors for the sustainability of food systems. Because the duALIne project has not considered the impacts of primary agricultural production (dealt with elsewhere by other foresight studies), greater emphasis has been placed on processing and distribution operations under the three pillars of sustainable development. Each food can thus be linked to a specific technological progression (an original combination of processes, each one generating properties) and to the organisational and qualification standards that are involved in these operations. Going against the perceived notion of uniformity, the diversity of foods is developed thanks to the emergence of new processes, the added value of which does not result in the marketing of products that are immediately identifiable as being new (*invisible technology*).

However, any analysis of the sustainability of these food systems raises different questions. In the present case, we highlight two levels of organisation that will enable an analysis of their complexity.

First of all, *these food systems are interconnected with energy and chemical systems within ecosystems* (Colonna, 2006). The food function is in competition with, and complementary to, other demands on these two systems (housing, transport, clothing, hygiene, etc.), which thus raises questions in terms of the allocation of resources, land and biomass (including water, fertilisers, pesticides, etc.).

Furthermore, the sustainability of food systems cannot be analysed without considering the diversity that already exists at the levels of product quality and the technical and organisational solutions that may differentiate these products at the production, processing and distribution stages. Food systems are composite entities that contain numerous 'subsystems', each with a specific coherence relative to the challenges of sustainability and quality.

These analytical approaches thus lead to the identification of new questions for research.

4.2   Interconnections between food systems and energy and chemical systems

Food systems are interconnected with energy and chemical systems within ecosystems.

The processing of agricultural materials into food involves several separate operations: (1) grinding, fractionation and purification operations which transform agricultural raw materials into first-generation intermediate food co-products (IFP), for use as either ingredients or additives, then (2) operations to enhance their technical and functional properties to obtain second generation IFP, (3) the formation of foods developed by assembling IFP with different ingredients and additives, and finally (4) packaging, which merits particular attention in that it can control not only the stability of foods but also purchasing and consumption behaviour. The consequence of the first step is a disappearance of the distinction between a natural product and a by-product to the benefit of co-products, each with their own functionalities and areas of application. This mainly concerns arable crop chains, the dairy chain (milk powder, cream, butter, whey proteins) and eggs (white and yolks). The production of animal feed consumes considerable quantities of agricultural raw materials and industrial co-products (oilseed cake, distillers' grains) and ranks as an adjustment subsystem in the conversion of solar energy into foods, bioenergies and bio-based products.

These different stages are present in all food systems, but are sometimes concealed to benefit an inappropriate and direct link between nature and food (cf. Figure 4.1).

4.2.1   The challenges

Analysing these interconnections enables the identification of several challenges for the coming decades:

- How can we achieve the required increase in biomass for *food* and the *renewable carbon industries*?

Greater global demand (development) is currently causing conflicts between yields, land and ecosystem services (the latter are currently being defined).

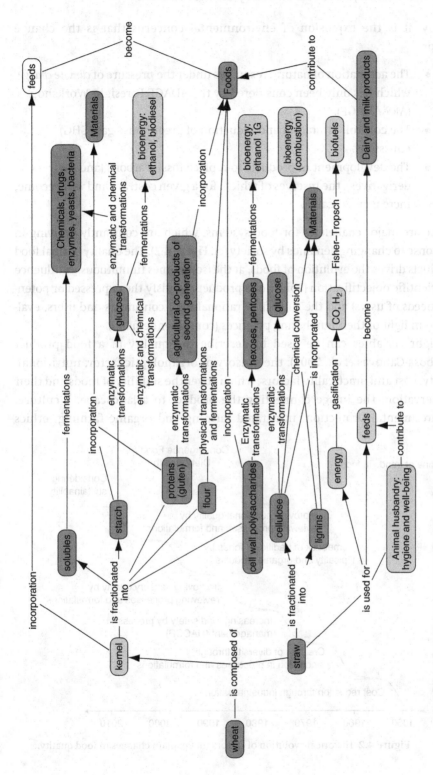

Figure 4.1 Role of wheat and its anatomical fractions in energy systems, food systems and chemical systems.

Today, it is the explosion of environmental concerns that is the change factor.

- The adaptation of natural resources under the pressure of *climate change*, which has only been considered by the ADAGE Foresight Workshop (ARP ADAGE).
- The control, limitation and reduction of greenhouse gas (GHG) emissions.
- The development of products to replace fossil carbons (and their derivatives), the reserves of which, for a given cost, are and will become, increasingly scarce.

These are major challenges for food systems, which are constantly evolving in response to changing demands by society (cf. Figure 4.2). The quality of final food products drives the evolution of foods, at the (sometimes tumultuous) confluence of scientific objectification (ability of a product to satisfy the expressed or potential needs of users) and the subjective rationality of consumers and users, evaluated in light of their purchasing practices (consumer demand).

Eight variables can be used to describe the quality of a food product (Barbosa-Canovas *et al.*, 2009): the sensory or organoleptic factor; nutritional, safety, cost and practicality factors, which cover the stability of foods and their preservation; the image connecting the product to a landscape, a culture, environmental protection, integrated farming and organic farming, ethics

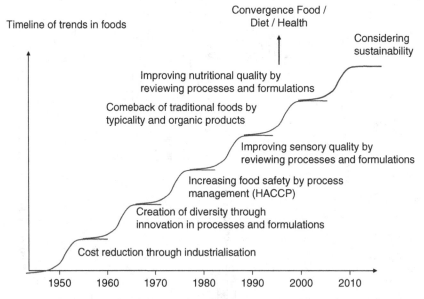

Figure 4.2 Historical evolution of factors that explain changes in food quality.

(with the development of fair trade products, such as the Max Havelaar® label). The evolution of all foods during the past century has been marked by successive additions of these factors (Figure 4.2). The rise of environmental concerns is the most recent evolution underway and requires a review of this concept so that it can be broadened to include the different components of sustainable development in an optimised manner.

The first specificity of the variables is that the *sum of the impacts of each stage between production and consumption* is only observed with respect to cost, health safety and image; in the latter case, it can also be associated with a landscape, culture or ethical approach.

By contrast, the other variables are *emerging properties*: the sensory, nutritional and practicality components are strongly determined by the formulation and final cooking stages, and cannot be deduced from the biochemical composition alone. The different levels of organisation of each food, and the conditions for consumption, determine sensory perception (through time-intensity responses) and the fate of the food in the digestive tract: the different mechanisms underlying the breakdown of the food and its biochemical components govern the kinetics of nutrient release and the fate of often non-assimilable and fermentable fractions. They explain the role of reverse engineering, which has been implemented to reduce salt, sugar and fat contents in recent years, while leading the *positive evolution of sensory standards. Reverse engineering places finished food products* in their consumption context, at the initiation of design and action, unlike valorisation, which is more appropriate to surplus-driven and monopolistic farming. The aim is to identify and understand the technological processes that generate both foods and products for non-food uses. So, a product will no longer be considered as having a structure and properties that were intended to result from a single pathway, but which on the contrary will arise from a choice from different (bio)technical options that respect the rules of sustainable development. The process thus becomes a generator of properties. Reverse engineering therefore combines an analysis of the specifications (a qualitative objective with non-compensable variables) with an overall objective of operational functionality: this combination leads to the association of heterogeneous constraints and their propagation throughout the process.

Thus the concept applied can always be summarised as follows. Based on biochemical food components, their ultrastructural organisation in agricultural raw materials and their intrinsic physicochemical properties, which processes need to be implemented, which structures and reactions need to be generated, to obtain the final properties desired? The final choice made by a socioeconomic, cultural or political decision-maker comprises a series of criteria, which are not necessarily known today. This approach, based on indicators that are characteristic of technological systems, enables access to the diagnosis, monitoring and prediction of the effects of any biological variations.

The second specificity is the *non-compensable nature of the characteristics of a food*. Because of this, quality results less from a consensus than from a managed and coordinated compromise, the input variables of which are the characteristics of the components, the type of processes implemented and how they are managed. It is therefore necessary to gain an understanding of the mechanisms involved during food processing, both in terms of implementing the process and choosing the raw materials.

4.2.2   Systemic approach

The biomass produced meets the different needs of humans, such as nourishment, clothing, materials, hygiene, etc., whose relative importance varies depending on the level of development and the sustainability choices made by each country (cf. Figure 4.3). Systems based on fossil carbon (oil, coal, gas) provide technical solutions that complement those resulting from biomass, except in the case of food systems at present. In developing countries, because of the use of wood energy, biomass increases the pressure on natural – and particularly forested – environments.

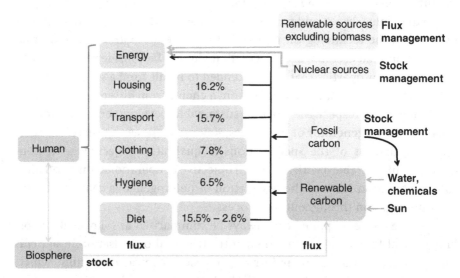

Figure 4.3 Overlap of food, chemical and energy systems in order to meet human needs. Numerical values indicate the relative share of French household expenditure by consumption type in 2006 (Source: INSEE, 2006).

The result of this analysis is that the choice of scale considered (regional or global) is crucial to our understanding of productive systems in the broadest sense. In the past, technological choices resulted from empirical know-how and then gradually developed through knowledge associated with decision-making methods, including on-line controls. At present, the methodology of experimental designs can answer a large number of questions without knowing or understanding the phenomena at play. The problem encountered when attempting to generalise this methodology resides in *the often non-linear interactions between the different steps* implemented, and in the large number of input variables. It is therefore necessary to understand and model all the interactions that occur at the different levels of combined structures, each corresponding to one or more phenomena at a given spatial and temporal scale. *A formal simulation framework* to enable the reasoning and formalisation of knowledge, and to break down and analyse the complicated interactions at play, will then make it possible to use *ASPEN plus*™ (Aspen Technologies, Inc., USA) or *Prosim* process simulation software to test the different scenarios.

A food system is therefore a conceptually rich entity that extends beyond the simple operations carried out in a food chain, broadened to cover all the consequences of technical choices in terms of substitutions, and generating scenarios for long-term and large-scale effects. Furthermore, the reciprocal interactions between social norms (relationships between people) and effects on the environment and resources are poorly understood.

However, the current trend towards modelling systems at a global scale cannot be immediately relevant to describing these overlapping relationships, at least in terms of mass and energy flows. The laws governing a change of scale from a farm or a company for analyses at the scale of a region, is an area that is still little explored. By determining these laws, it would be possible to prevent the enormous uncertainties observed at the output of global models. *Theoretical foundations and tools that can integrate biological/physical/chemical/economic processes in indicators and models at different scales (of time, space and organisation), and the laws which govern a change of scale, are of crucial importance. Combining temporal dynamics (year for production, week for processing, day for distribution) with adaptable storage functions is also a major challenge* (Minegishi and Thiel, 2000; van der Vorst et al., 2000) (cf. Figure 4.4).

It should finally be emphasised that food production + processing + distribution is based on the use of different types of energy (greenhouses, freezing, cooking) or products for which energy consumption is necessary in order to obtain them (fertilisers, equipment). Hence we need to reconsider the technologies involved (fractionation/recombination, high temperatures/freezing, dehydration/rehydration, etc.) using validated data.

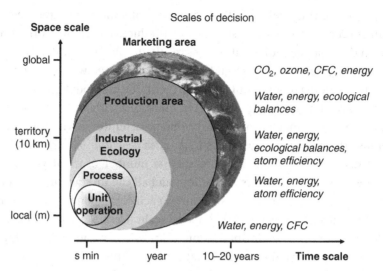

Figure 4.4 Decision-making scales concerning critical variables for the sustainable development of a food system.

Questions for research

It is recommended that research returns to fundamental issues, with the rigorous establishment of energy and mass balances (renewable carbon, phosphorus and nitrogen) for each stage of the operation. Then, by applying combinations of operations, and using different agricultural raw materials, a capacity may be developed to simulate different technological trajectories.

The challenge is to be able to describe the trajectories for the optimum sustainability of the circular economies of renewable carbon (Gray, 2009; Lundqvist *et al.*, 2008; Muñoz *et al.*, 2008) and nitrogen, and to a lesser extent phosphorus, with spatial and temporal differences. Uncertainty, which arises from the variability of raw materials and the range of parametric choices for each of the processes, leads to considerable overall uncertainty. Its reduction requires the integration of data, the know-how of experts and specific knowledge, exploited at the different scales and in different ways by the group of actors involved in the same phenomenon. Methodological research is therefore necessary to determine the entire technological flow chart that will allow the management of raw materials required for converting the latter into the product family (or families) targeted (viability tube) and the associated strategies, as well as the robustness of these strategies.

The quantification of uncertainty poses the problem of the implementation of uncertainties in systems, as well as their propagation throughout the final evaluation process. This point is essential to facilitating decision-making and identifying the most critical obstacles in a system. A reduction of impacts can only be analysed by reducing flows of mass and energy throughout the system, and not by applying a minimisation at each stage. The difficulty lies in understanding the interactions between different components in the system.

Contrary to the commonly accepted opposition between bioenergy and food, the 'Bioenergy and Food Security (BEFS) Analytical Framework' (FAO, 2010a) seems to have been the first initiative to have followed the path towards reconciling these two end uses in the countries it considered (Peru, Thailand, Tanzania).

4.2.3   Global systems with different purposes, to be considered at a regional level

The management of food/energy/chemical systems can have different objectives of a socioeconomic and/or geopolitical nature:

- to encourage energy independence
- to guarantee food sovereignty
- to initiate carbon-neutral development (circular economy)
- to develop a bio/agri-industry that creates added value, etc.

The ecosystem unit that should be considered when developing a renewable carbon system must comply with two characteristics: (1) it should be shared at a global level, to facilitate extrapolation to this scale, (2) it should reflect a complete unit with a balance between climate (temperature and rainfall) and land, carbon stocks and plant substrates, so as to determine the plants that can be planted in the biome and their capacities for primary biomass production. *Consequently, the biome (Prentice et al., 1992) is the correct scale at which to perform modelling, linking ecosystem resources and their potential uses in a context of sustainability.* The impact of climate change can be monitored at this scale. The limitation of the models available at present (Sieber, 2006) concerns an evaluation of the availability of sufficient volumes of water for agricultural purposes.

Current studies on food systems do not take account of the limited capacities of each biome to meet human requirements. The sustainability analysis could be repeated, starting from each of the different biomes determined and which can be considered near to environmental equilibrium. Indeed, this is reflected by the recent proposal concerning the development of national rather than European sustainability criteria made by Sweden, Austria and Finland and three Baltic States. Unlike standard attributional LCA, the consequential LCA that are currently being developed will generate scenarios that reflect long-term and large-scale effects involving exchanges beyond the levels of biomes (Earles and Halog, 2011) and social norms (relationships between people) when they take account of environmental effects. The industrial ecology approach (Adoue,

2007) is able to stretch beyond sectoral environmental policies that reason in terms of waste reduction at both sides of the pipe to lead to global optimisation of resource use. The French Pomacles Industrial Zone is an illustration of this approach; it was set up for sugar production and has since shifted towards the development of a cluster of companies which exchange the products that arise from the fractionation of different agricultural raw materials (cereals, sugar beet, etc.) and their by-products, including fermentation CO_2.

Evaluation is often neglected when it comes to the environment (see Chapter 10 on Methods). It frequently tends to focus on environments that are little affected by human activities, with particular emphasis on biodiversity and health risks. The problems involved in environmental evaluations are numerous:

- the multi-criteria nature of the analysis;
- spatial and temporal uncertainties, a long timeframe and how these parameters are taken into account (how inaccuracy affects outputs);
- the ranking of environmental criteria relative to the two other pillars of sustainable development, notably in Southern countries.

If land use, biodiversity and water resources have to be taken into account, specific indicators must be developed.

LCA is a tool that can only be applied to each food (functional unit). The scientific challenge is the eco-design of food systems, which requires a systemic approach using detailed modelling and reliable databases. The eco-design of foods is pertinent in a field in which there are considerable possibilities of substitution at the formulation level. However, it must be ensured that the eco-design of foods is not a second-rate factor, compared with the role of animal products which are certainly a first-rate factor in the sustainability of food systems (Haberl *et al.*, 2011; Lang *et al.*, 2011).

4.2.4   *Towards the design of biorefineries*

The updating of systems that involve the eco-design of dedicated pro-duction/processing systems so as to target a circular economy rather than a substitution logic, means that a central role must be given to biorefining that is fed by both plant products and agri-industrial and household waste. The growth of urbanisation, and more particularly of mega-cities, requires considerable and concentrated management of the waste generated, which constitutes a source of biological carbon. The context is thus favourable for the development of biorefineries.

This only constitutes an update of step 1 of the processing of plant materials, with an extension towards bioenergies and green chemistry. The prospects

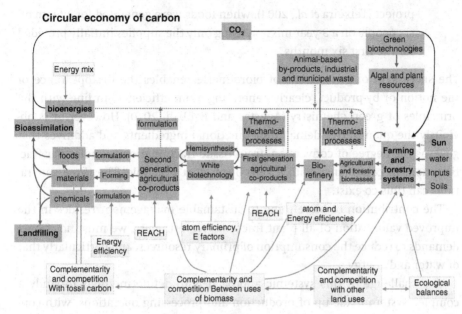

Figure 4.5 From challenges to drivers.

offered by white biotechnologies (microbiological) and green biotechnologies (plant-based) will enable greater use of plant fractions, and particularly of lignocellulose, although the return of biomass to the soil after harvest should not be neglected (cf. Figure 4.5).

The development of biotechnologies should be accompanied by an evaluation of their added value at each stage.

Over and above semantic considerations, three generations can be seen:

- **Generation I**, with the fractionation of tissues (milling, oilseed grinding, sugar production, etc.). The flexibility of the raw materials used, process diversity and by-products are all reduced.
- **Generation II**, where subcellular fractionation, combined with chemical and biological conversion, are based on a greater diversity and flexibility of processes, and result in a diversity of by-products (starch milling, lipid chemistry, dairy production).
- **Generation III**, molecular fractionation down to monomers to obtain energy and molecules. The intensive use of bioconversion leads to greater flexibility with respect to raw materials, processes and by-products. Starch, lignocellulose, food and household waste, all constitute resources in this context. Micro-algae and yeasts are important players, enabling the modification of carbon motifs. This logic was pushed to its extreme during the NASA Advanced Life Support

project (Teixeira *et al.*, 2004), when foods were generated for a team of six people on a 3-year mission using only the supplies initially provided for the first six months.

The succession of generations of biorefineries enables the disappearance of the notion of by-product, clearly generating atom efficiency, in line with the principles of green chemistry (Anastas and Eghbali, 2010). However, it is obviously the quantitative demand for functional ingredients and additives that governs the degree of conversion, from tissue fractionation to molecular fractionation, in compliance with a zero by-product principle. The three generations can thus co-exist.

The contribution of biorefining to sustainable management resides in the improved valorisation of all plant fractions. Nevertheless, we must not forget demands to reduce the consumption of primary resources, and particularly that of water and energy.

The challenge of this systemic approach is to define methods to study a complex system made up of production and processing operations, with considerable human intervention throughout. The important question is to determine how it is possible to obtain the small amount of information that will be determinative in describing system state variables. In this context, the development of experimental strategies based on an understanding of the operations involved is essential and determinative. The corollary is a need to assemble and integrate knowledge so as to facilitate operational choices that will reconcile the three challenges of food, energy and chemistry in a sustainable biosphere.

Thus food systems are only one element in more global ecosystems. Furthermore, they themselves do not appear to be homogeneous, but are made up of a diversity of subsystems that can be detected in light of the consistencies established between consumption, exchange and production modes. The second part of this chapter is devoted to studying the internal dynamics of food systems.

4.3   The diversity of food systems: evolution and challenges for sustainability

Different types (or ideal types) of food systems can thus be identified. This classification is detailed below, together with the variables upon which it is based (Section 4.3.1), and the sustainability of some 'emblematic' alternative systems (such as Geographical Indication products, organic farming, fair trade) is analysed (Section 4.3.2), before considering the outcomes of the observation that these ideal types do not function in isolation but within a permanent process of interactions, recombination and hybridisation (Section 4.3.3).

4.3.1   Classification of food systems

4.3.1.1   Differentiation variables

The characteristics of food systems can be clarified using an analytical approach to identify their underlying structure. The characterisation of food systems thus supposes the determination of pertinent variables, which may belong to one of three groups:

- Structural variables, generally applied in *food chain* and *supply chain* analysis:
  - (a) geographical proximity between activities, or between production and consumption (sometimes simplified to the notion of food miles): enables a distinction between local and 'long-distance' food systems
  - (b) number of intermediaries: enables a distinction between short and long supply chains
  - (c) importance of processing and incorporated services: 'raw' versus 'processed' products with agricultural, service or processing components
  - (d) type of product relative to its uses: this takes account of its perishability, weight, volume, natural production conditions, food versus non-food uses, nutritional values, etc.
  - (e) proliferation versus concentration of functional/economic units at each stage
  - (f) consumption sites: at home or away from home
  - (g) storage methods for foods and ingredients at different stages.
- Variables emphasised in more *political (or institutional) approaches* to sectors:
  - (h) organisation of work at different stages (production, processing, distribution, consumption): home versus salaried activities; small scale versus industrial; capitalist versus non-profit orientation, etc.
  - (i) competitive and trade context: degree of concentration (monopolies, oligopolies, competition, etc.), type of regulation mechanisms, level of financialisation of trade, risk management methods, etc.
  - (j) role and forms of public action: intervention by States, local government bodies, international institutions, etc., and the legitimacy of this intervention, control of the system, direct or indirect support, etc.

      (k) international integration: importance of imports and exports, role of global corporations, international institutions and agreements, role of private networks (cf. global value chain)

      (l) governance: control of decision-making and information, standardisation (approaches via contracts and value chains).

- Variables considered through *cognitive approaches* to food systems, including the challenges of knowledge construction, common values, agreements on key components in the system, etc.:

      (m) product quality: by specifying the attributes defining product quality, the agreements that stabilise this quality (agreements on values underlying a particular quality), standards/signs and organisations with which quality is associated

      (n) territorial relationships: agreements, representation of links between products, activities and geographical areas

      (o) role and legitimacy of technologies (the GM/non-GM debate is an example with a high media profile in Europe)

      (p) social considerations: challenge of social justice in production, trade, consumption; ethical, cultural or religious dimensions

      (q) knowledge and control of culinary know-how regarding food preparation.

Variables n, o, p and q may be partially integrated in variable m on product quality.

4.3.1.2   The different types of systems

Taking account of these variables enables the construction of a classification that distinguishes five food system ideal types with several variants:

Domestic food systems

In this type of food system, consumption mainly occurs at the level of the production unit, where unconsumed (excess) products are sold locally or in the context of non-market exchanges based on family or social links. Processing is also domestic, or may include some types of cooperative organisations. The functional mode is individual or family-based, or may be organised by collective groups (family-owned or shared gardens). This type of food system ensures the subsistence of many populations in Southern countries, but can also be developed in both Northern and Southern countries coupled with other supply modes.

Local food systems

These food systems comprise a small number of intermediaries (short chains, direct sale, etc.) and there is (generally) geographical proximity between producers and consumers. Quality is developed through knowledge (direct or indirect) of the producer by consumers, so it does not necessarily require any official signs of quality. This type of food system, historically present throughout the world, is currently regaining strength in Northern countries thanks to the involvement of local government bodies (e.g. local markets), consumer associations or producer organisations. The practical organisation of chains may therefore take different forms. These local food systems are sometimes coupled with efforts to ensure health or gustatory quality, while proximity promotes quality attributes that are difficult to measure.

Regional food-producing systems

Basic food products (cereals, tubers, pulses, fruits and vegetables) and other food products that can be stored, travel moderate distances (100 to 1000 km) between the regions of production and consumption (the latter often being urban). As a general rule, the chains involve collectors, wholesalers, small-scale or semi-industrial processing companies (micro-businesses or small businesses) and retailers (with transport by lorry or rail). Few contracts are made between players, and relationships tend to function more on the basis of personal and trust relationships.

This type of system exists traditionally in both Northern and Southern countries. In the latter case, however, the chains are markedly restricted by transport problems (particularly in Africa) and by the instability of production and consumption (due to price fluctuations). The relationships between players are also more personal, sometimes involving networks led by urban-based wholesalers who maintain credit relations with their suppliers and play an informative role. The principal logic (or consistency) underlying these networks is the security of supplies and redistribution, the diversification of activities, small material investments (generally leading to a relatively low overall cost) and flexibility (Hugon, 1985; Moustier et al., 2002).

Agri-industrial (or 'tertiarised agri-industrial') food systems

The principal objective is to produce food for mass markets while reducing costs and maximising profits. This type of food system is based on the development of technological or commercial economies of scale, and on processes for specialised production within producing areas. It has given rise to the growth of processed products and incorporated services. The economic, geographical and cognitive distance between producers and consumers is considerable, with a relatively large number of intermediaries (collection centres, industry, purchasing platforms, distribution). Standardisation and regularity are the attributes that underlie product quality, enabling compliance with national and international regulations in terms of health, social and environmental requirements, etc. This type of system has also been developed under pressure from consumer movements focusing on food systems, mainly with respect to controlling price levels and establishing global systems for information and guarantees (in terms of quality, traceability, health or nutrition) (Dubuisson-Quellier, 2009a).

Differentiated quality food systems

The underlying logic in this case is the differentiation of products and foods justifying a higher unit cost and (generally) a higher final cost as well. Labels and signs of quality guarantee product attributes that cannot usually be assessed upon purchase. This type of food system can only become stable if consistent quality management exists throughout the chain (in terms of information and controls) and if recognition by the consumer generates sufficient income to cover the additional costs of production/transport. Different subsystems are linked to the selected quality attributes, which are guaranteed by quality labels or private brand names:

- Food systems with *heritage* qualities; e.g. products with Geographical Indication status. Quality is associated with the origin of the product, local production conditions and a related heritage (nature, landscape, know-how, culture). This supposes that the consumer is attached to the region of production and, if possible, recognises specific intrinsic characteristics. The chains are generally long, but coupled with tourism, and these products are often exported. Public intervention may be necessary to organise producers and provide guaranteed resources, but not results.
- Food systems with *natural* qualities; e.g. organic products. Quality is based on the recognition of 'environmentally friendly' practices or particular 'links with nature'. Most chains are long but remain diverse, as do organisational modes (familial, salaried employees). These systems may incorporate a broad variety of values for consumers: some may place greater emphasis on protecting the environment (Germany) or on food safety (no use of synthetic plant health products (France)).
- Food systems with *ethical*, religious or community qualities; e.g. fair trade, kosher, halal, etc. Quality derives from practices and organisations that respect moral or religious values, and/or from solidarity with a particular population (small producers in Southern countries, disabled workers, religious groups, etc.). This quality is guaranteed by special labels or private brands.

- Food systems with *superior gustatory qualities*; e.g. 'Red Label' products. This superior quality is linked to specific varieties or breeds, technologies or know-how, and is not necessarily connected to a particular geographical region.

4.3.1.3   Assessment

If the purpose is to analyse sustainability, these different food systems will need to be studied in more detail.

The predominance of the agri-industrial model seems to be confirmed in both Northern and Southern countries: it now accounts for between 40% and 45% of world food consumption, according to the estimates produced by Rastoin and Ghersi (2010), and is still seeing strong growth in Southern countries (particularly South-East Asia, Latin America and Southern Africa).

However, this model is not in the majority and all other concomitant systems must also be considered. 'Traditional' systems (domestic, local, regional, and some components of differentiated quality food systems), marked by a strong artisan dimension, still hold a global majority with between 50% and 55% of world food consumption (Rastoin and Ghersi, 2010).

These traditional systems remain the principal source of supply for many populations in Southern countries. They have demonstrated their resilience in a context of significant climate and market instability. They have long been neglected by government support, but new aid opportunities are now developing, more suited to the decentralisation and labour-intensive, capital-extensive characteristics of these systems: gradual development of producer organisations, professional associations, ecological farming practices, improvements to transport, storage and communication infrastructures, savings and loans schemes (de Schutter, 2008, 2011). Research has an important role to play in characterising trends in these systems, and the comparative efficiency of different production and distribution systems, in order to combat food insecurity. Indeed, in Southern countries, such situations of food insecurity are poorly documented because of the constraints regarding the implementation of observations of food consumption.

In Northern countries, where the agri-industrial model has been developing for longer, these traditional systems have been maintained (or (re)developed in some cases) as an 'alternative' to the agri-industrial system, and demand greater consideration of the challenges of sustainability (local, farm-produced, organic, regional and fair-trade products corresponding to 'local' or 'differentiated quality' food systems).

All five food system types in this classification should thus be considered because they relate to different *consistencies established between how products are produced/processed/distributed/consumed* and to different ways of constructing each of the product's qualitative characteristics. The players in these systems adopt markedly differing positions relative to sustainability challenges; implicitly or explicitly, they determine priorities among the different dimensions (economic, social or environmental) of sustainable development. However, none of them represent a global approach to the entire food system, and even less its applicability to the whole world.

4.3.2   Sustainability of food systems

4.3.2.1   Contributions and limitations of the agri-industrial model

The sustainability of the agri-industrial model has been analysed by numerous authors (Rastoin and Ghersi, 2010, for a review of this issue). This type of intensive, specialised, concentrated, financiarised and increasingly globalised system has many positive aspects that have distanced the threat of famines, caused a very marked drop in food prices, ensured food safety, maintained very strong economic activity and offered hyperchoice to consumers in industrialised countries; however, this model has not proved its ability to feed the planet and, in parallel, obesity and other diet-related diseases are affecting an increasing number of people. The negative externalities generated by this model are also increasingly being documented:

> [the] exhaustion of natural resources and degradation of landscapes,
> hyperspecialisation of production units and an artificial
> hyper-segmentation of products that aggravates economic disparities
> between companies and between consumers [. . .], the relocation of
> activities to sites that benefit from comparative costs and from which
> products are exported throughout the world [. . .], destroying small local
> producers [. . .] and, in the longer term [causing] the disappearance of
> a regional culinary heritage [. . .] and increasing vulnerability to pandemics.
> (Rastoin et al., 2010)

Based on these observations, many stakeholders (consumers, producers, governments and civilian society) call for the maintenance of older types of food systems, or the development of new solutions, usually combined with, or complementary to, the agri-industrial system. These 'alternatives' to the agri-industrial system have thus rehabilitated older types of food system (local or differentiated quality systems, such as regional products), or led to attempts to develop new types of quality differentiation, new labelling initiatives for production/marketing systems for agricultural and food products, such as organic farming or fair trade.

An analysis of four types of 'alternative' food systems (in which we include both older and innovative systems) chosen as the most emblematic – short chains ('local' food systems), products with Geographical Indication status, organic farming and fair trade ('differentiated quality' food systems) – will allow us to make an initial assessment of the conditions for their emergence and their economic, social and environmental impacts.[1] This arbitrary choice should in no way conceal the need (referred to above) to explore other types of food systems (domestic or regional). It should also be remembered that all these 'alternatives' are still under debate regarding their ability to meet all the needs of our planet.

4.3.2.2   Short chains and the call for a 'food democracy'

A whole series of initiatives have developed in recent years, focused on establishing more direct relationships between agricultural production and food consumption, or between agricultural producers and consumers, seeking thus to restore older types of more local food systems. It is now necessary for local consumer organisations, producer and consumer groups or regional government bodies to ensure the practical organisation of food supply systems whose control is no longer delegated to third parties but is exercised by committed participants.

A review of the English language social sciences literature devoted to analysing these alternative systems (Deverre and Lamine, 2010) highlights the fact that although few of them explicitly claim to derive from sustainable development, their position relative to the 'dominant' system is based on criticism of the supposed environmental impasses of the latter, its lack of equity (reserving high-quality products for the most wealthy consumers) and a choice of producers that excludes many of those with the least capital or technical resources or who are the most exposed to competition from countries with low labour costs. Depending on the country, the priorities claimed by the promoters of these systems vary (access to healthy food for under-privileged populations, response to sanitary crises attributed to the dominant system, defence or promotion of 'peasant' or 'organic' production methods and consumption modes that are better balanced and respect seasonal cycles, etc.). This variety is reflected by that of the practical organisation of these systems, ranging from those with the highest media profiles such as consumer–producer collectives (Japanese teikei, Community Supported Agriculture in the USA or UK, AMAP in France) or the restoration of country

---

[1] Concerning this question of the sustainability of local and differentiated quality food systems, see also Chapter 6.

markets in countries from which they have disappeared, to organisations developed at other scales, such as Food Policy Councils which aim to control food supplies at the level of regional governments, for example through their use by collective catering or food aid for the most underprivileged.[2]

Although the great majority (if not all) of these systems are based on more direct and local relationships between agricultural production and food consumption, their reduction to the notion of 'short chains' (as is currently the case in France regarding the French Rural Network) generally conceals the strategic political dimension they represent, for their promoters if not for all those committed to them. Terms such as 'food citizen', or more commonly 'food democracy', illustrate this desire to politicise the issue of food by associating groups of producers and end users at the level of both small communities and elected local government bodies. For consumers and/or producers, the aim is to regain collective decision-making power within food systems and no longer to delegate this power to market operators or administrative structures for regulation and control.

The promoters of short chains thus seek to develop more sustainable food systems, mainly by acting on what is presented as the fourth pillar of sustainable development, i.e. the governance of these systems. They then appear to implicitly rank the other pillars, placing the social dimension ahead of economic and environmental dimensions.

However, the impact of short chains on these different pillars remains a matter of debate. For example, some critical analyses have emphasised:

- the lack of strong evidence for the advantages of proximity in terms of energy consumption or greenhouse gas emissions;
- social selection, which operates despite the desire to respond to the nutritional problems of underprivileged populations;
- the risk that farmers may become subordinate to urban groups in new ways;
- regional inequalities, which will inevitably cause the relocation of food supplies, or even the development of certain types of protectionism or 'food racism'.

At an economic level, the ability of this distribution mode to meet quantitative food needs is also hotly debated.

[2] A more ambitious theory was put forward by Kloppenburg (Kloppenburg et al., 1996) regarding the notion of the 'foodshed' of a community or food consumer group. This notion has been taken up by some collective groups committed in English-speaking countries to the *Transition Towns* movement, which includes the issue of food supplies in their Agendas 21.

4.3.2.3 The territorial roots of food products and the Geographical Indication system

Because of the increasing industrialisation and globalisation of trade, the protection and development of 'regional' products and 'traditional' production methods has necessitated the implementation of certification systems.

In 1883, the Paris Convention laid down the regulations for the national registration of collective brands or certifications which would indicate that the products concerned benefited from specific qualities (which could include their geographical origin).

The start of the twentieth century saw the development of a series of systems enabling the legal protection of regional products: *Appellations d'Origine* (AO) were recognised in France in 1905. These arrangements then spread worldwide, with the recognition of Geographical Indications (GI) as an intellectual property right in the context of the Agreement on Trade-related Aspects of Intellectual Property Rights (TRIPs), signed in the framework of the World Trade Organisation (WTO) in 1994 (AFD-FFEM, 2010; Vandecandelaere *et al.*, 2010).

Geographical Indications could be viewed as alternatives to an agri-industrial production mode focused on intensification and standardisation, and for this reason as an element that might enhance the sustainability of production systems by allowing regional development that respects natural resources and has a positive social impact. However, although the specifications for GIs must demonstrate a link between the typicity of a product and its production region, they are not required to guarantee an environmentally friendly production mode or any type of social equity (see the work carried out in the context of the *SinerGI* research programme and on 'non-sustainable' GI tequila; Bowen and Valenzuela Zapata, 2008; Vandecandelaere *et al.*, 2010). Although GIs are granted in order to valorise a specific local raw material, they may however enable better remuneration for the producers supplying these raw materials (for example, the case of Beaufort cheese compared with other GI cheeses; see Sylvander *et al.*, 2006).

Consumers and citizens are now becoming increasingly interested in regional characteristics, knowledge of production methods or products linked to local history. GI products may convey a food culture and links between a region and food. They thus constitute a link to the cultural and food identity of a region and country.

The impact of these certification systems on the origin of 'regional' products remains conditional upon the efficiency of the legal system that protects them. At an international level, TRIPs can only ensure minimal protection, except for wines and spirits that benefit from additional guarantees.

This impact also remains conditional upon the existence of local policies to deploy these protection systems in favour of producers and territorial development. Expectations regarding GIs are highly diversified, in both Southern countries (where rapid development of GIs is being observed) and Northern countries. In many cases, the principal aim of registering a GI is to open up new export or domestic markets and to ensure the protection or reinforcement (or even creation) of a reputation, and not necessarily to provide a community of producers with the means of enhancing their territorial resources or defending or developing a rural area.

Geographical Indications may have the potential to reinforce the sustainability of food systems, mainly with respect to social aspects (territorial development, including that of marginalised areas, encouragement of heritage value/protection of local resources, etc.) and economic aspects (better remuneration of producers, changes to the distribution of added value, etc.). However, it is important to pursue analysis of this potential at environmental, socioeconomic and political levels.

Many countries are now committed to the recognition and protection of their regional products thanks to GIs and WTO regulations. It is necessary to analyse the ability of these regulations to create markets for specific raw materials that promote biodiversity and allow higher prices, notably for the producers of raw materials. In other words, GIs need to become a tool for agricultural and food policies.

4.3.2.4   The development of organic farming and its economic, social and environmental impacts

Organic farming is quite 'naturally' classified within sustainable agriculture, for instance by the French government (cf. conclusions of the Environment Round Table[3]), by the EU (which includes it in its sustainable development strategy), and by the FAO (which sees it as an element to reinforce food security), amongst others. However, the impact of organic farming on the three pillars of sustainable development is questioned in the scientific literature.

At the environmental level, many studies have highlighted a positive impact, arising from both the non-use of synthetic chemical products and the development of rotation plans for land use. However, the issue of greenhouse gas emissions remains controversial, as few studies have included an evaluation of transport (Capitaine et al., 2009).

Indeed, there exists real diversity among organic farming development models (Desclaux et al., 2009; Sylvander et al., 2006). Are studies that compare a single

---

[3] Known as 'Grenelle de l'environnement'.

organic model with conventional agriculture (itself considered in terms of only one model) of any real value (Capitaine *et al.*, 2009)? More generally, some authors have underlined the incompleteness of impact studies that have focused on particular chains and neglected certain aspects (ecological services and biodiversity) (Blackman and Rivera, 2010). Finally, recent years have seen the growth of assessment models, resembling LCA techniques to a varying extent:[4] by adopting different scales, or considering different types of impact, they have reached different conclusions. LCAs themselves can produce contrasting results, depending on the functional unit chosen (Basset-Mens and van der Werf, 2004; see Chapter 10).

Few studies have been performed on the social impact of organic farming. Although it is often proposed that organic farming includes 'an ethical dimension which, according to the International Federation of Organic Farming Movements (IFOAM) is reflected by [...] social and humanist objectives (international solidarity, closer ties between producers and consumers, cooperation and non-competition, equity between all stakeholders, maintenance of producers on the land, safeguard of rural employment, etc.)' (Bellon *et al.*, 2000), few studies have demonstrated this. The literature contains several critical analyses on this point, listing the negative impacts. Thus Getz and Shreck (Getz and Shreck, 2006) noted, in the case of organic tomato producers in Mexico, that organic certification had 'exacerbated socio-economic inequality and disrupted local social norms' by placing too great an emphasis on the need for control and mutual surveillance. This refers to the limitations of the Third Party Certification system, to which the IFOAM has responded by developing participative guarantee systems. However, according to other authors, the implementation of organic certification is likely to sustain social capital among producers (Bray *et al.*, 2002).

The economic profitability of organic farming is clearly dependent on the yields it can achieve. Comparing the yields of organic and conventional farming remains a controversial issue (and is rendered difficult by the numerous models available, as mentioned above) (Halweil, 2006).

---

[4] IDEA (Indicateurs de Durabilité des Exploitations agricoles, or indicators of the sustainability of farms), INDIGO (INdicateurs de DIagnostic GlObal à la parcelle, or indicators for a global plot diagnosis), DIALECTE (Diagnostic Agro-environnemental Liant Environnement et Contrat Territorial d'Exploitation, or agri-environmental diagnosis linking the environment and regional exploitation contracts), DIAGE (DIagnostic AGro-Environnemental, or agri-environmental diagnosis), ARBRE (ARBRE de l'exploitation agricole durable, or tree of sustainable farm management), PLANETE (Pour L'ANalyse EnergéTique de l'Exploitation, or for the energy analysis of farms), etc. Life Cycle Assessment (LCA) remains the only method that is standardised at an international level (ISO 14040).

It also reflects consumer demands. There is a propensity for consumers to be willing to pay for more expensive organic products, but only on certain conditions: they must also find other motivations, such as the 'health' argument (Sirieix *et al.*, 2006) and organic certification, a belief attribute, must be combined with other, more verifiable attributes based on knowledge or experience (Grolleau and Caswell, 2006). In this context, the demands of these consumers may force changes to the regulations by imposing obligatory yields on organic production, which at present is only subject to a best effort requirement.

However, this acceptance of some consumers to pay higher prices for organic products will only guarantee economic profitability based on the assumption of a certain efficiency of its production sectors and a good distribution of added value. This hypothesis can however be called into question: in the case of small-scale Indonesian coffee producers, Neilson (2008) anticipated a fall in 'farm gate' prices (paid to the producer) following the rise in transaction costs in the sector that could result from certification mechanisms. Once again, Third Party Certification is called into question to a greater degree than organic farming itself.

Finally, from the consumer standpoint, the 'additional cost' of organic products may make access difficult for some groups in the population. However, the promoters of organic farming respond to this argument by saying that, subject to an alteration of diets (more fruits and vegetables and less meat), organic products are not more expensive.

The sustainability of organic farming, based on its three pillars, thus remains a fully fledged research theme. This issue has been rendered more complex in recent years by the strong growth of organic production and its markets. This demand has risen most markedly in Northern countries, encouraged by changes in consumer expectations, in addition to a strong sustainable public procurement component (the supply of collective catering and school canteens), which in France followed the Environment Round Table.

This rise in demand has been accompanied by an uneven increase in production in different Northern countries. One result of this has been an increase in imports, so that the transport component weighs upon the carbon footprint of organic products. Organic production in Southern countries remains insufficiently documented to analyse the impact of this development (Blackman and Rivera, 2010).

Organic farming has developed worldwide, which has led some authors to question its ability to feed the planet. Its partisans consider that organic yields can equal those of conventional agriculture, or may even be higher. This theory does not concern the unit yields of particular crops but the global productivity of the system in terms of biomass and energy, including the effects of the rotations and combinations recommended under organic systems. However, they acknowledge that this is only possible several years after conversion, the period necessary for the soil to recover its natural capacities and for the farmer to acquire sufficient technical expertise. Yet the question of feeding the planet is not only dependent on the dynamics of technical innovations; its political and economic dimensions remain fundamental (Halweil, 2006). Indeed, some studies have expressed concern over a disappearance of the 'original spirit' of organic production (Conner, 2004).

4.3.2.5 Fair trade and a 'fair' price for producers

Fair trade aims to guarantee a 'fair' price for producers, enabling them to cover their production costs and make a sufficient profit to provide them with decent living conditions and a minimal capacity for investment. Above all, fair trade targets the social dimension of sustainable development, although minimising its ecological footprint is now increasingly being included in the objectives.

Reviews have recently been published on the impact of fair trade (Vagneron and Roquigny, 2010). They testify to relatively high capitalisation, but do not cover all stakeholders, chains, geographical regions and types of impact, and do not take account of long-term effects.

Different areas of debate have developed. According to their detractors, fair trade certifications focused upstream cannot guarantee a positive social impact throughout the chain (Jacquiau, 2006). This impact can only be ensured through the development of alternative chains, resolutely separate from conventional chains (as proposed by Minga, a French association defending a 'fair economy').

At present, fair trade involves encouraging greater responsibility and awareness among consumers. Although some authors consider that it is right and fair for consumers to be able to 'vote with their money' in favour of sustainability (Granatstein and Kupferman, 2008), others talk of over-responsibility and see a risk that this market will become 'the expression of society's projects' in place of public regulation (Figuié and Bricas, 2008).

The current 'scale change' of fair trade, in terms of its volumes and markets, has rendered these debates more virulent. Some feel that it is leading to a proliferation of brands, an increase in competition between producer organisations and the reintroduction of a market logic into chains through the presence of stakeholders with varying motivations (by comparison with the pioneering fair trade organisations) (Lemay et al., 2010). If a market rationale becomes a strong feature within fair trade chains, this could further reinforce economic inequalities among producers (Getz and Shreck, 2006).

For some authors, this situation confirms the limitations of the private regulation of fair trade; control based on a public/private tandem may be preferable (Raynolds *et al.*, 2007). But for others, it is the limitations of Third Party Certification that are highlighted (Neilson, 2008).

Above all, growth of this market requires the combined growth of organic and fair trade products (Seifu *et al.*, 2007). The benefits of fair trade have often been used to support conversion to organic farming. The need for a common strategy to be adopted by these two labels is confirmed by some authors (Gordon, 2005).

Thus fair trade is currently at a 'crossroads'. Lemay *et al.* (2010) see three possible scenarios for the future: (1) 'a multiplication of initiatives that will lead to a trivialisation of the effects of fair trade', (2) 'regulation mechanisms that will markedly institutionalise fair trade', and (3) 'fair trade will reorganise itself around new practices'.

Today, the ability of fair trade to drive changes in the global food system is questioned less and less. This point is indeed clearly included in the founding principles of fair trade ('Improve the rules and practices of conventional international trade'; the consensus view of FINE (2001)). In 2000, fair trade could be seen as an oppositional movement (Raynolds, 2000). Since then, the growing involvement of 'conventional' players, and the sector's massive growth, have relegated these debates to the background. The commitment of consumers is seen as a more complex issue, combining mechanisms of delegation through the choice of products with more direct types of commitment to alternative regulation (Dubuisson-Quellier, 2008).

4.3.2.6   Overview

Analyses of the development and impact of these alternative food systems, which are increasingly present in the scientific literature, have thus produced contrasting results. These alternative systems display considerable diversity, with each type focusing on different objectives and having developed equally diverse operating modes, including (or not) certification arrangements (determined either by a third party or under participative guarantee systems). The accreditation of these products or chains can be driven by different groups of stakeholders, depending on the country: national government, food production/distribution companies, producer or consumer associations, etc.

Measuring their impacts is rendered complex by this diversity, a lack of data, the proliferation of labels under different systems (for 'differentiated quality' food systems) and their frequent juxtaposition, or a lack of historical experience, etc.

Nevertheless, it seems quite clear that the key issue concerns the scale change that affects the development of these systems, which will lead to a reorganisation of this landscape.

4.3.3   Ongoing interactions, recombination and hybridisation

Food systems can indeed be seen as components of an agri-industrial system which is predominant in Northern countries and is seeing considerable expansion in Southern countries, around which gravitate different 'alternative' systems (a category in which we thus place traditional or more innovative food systems that claim greater sustainability, as discussed above). However, this general scheme remains purely analytical: a more dynamic analysis of this question shows that the development of alternative food systems that imply criticism of the agri-industrial model (Boltanski and Chiapello, 1999), based on the values of sustainable development, influences the latter via consumers, stakeholders or politicians.

The most innovative alternative food systems can be viewed as laboratories experimenting with technical and organisational solutions that have been adopted, or could be adopted in the future, by actors in more conventional systems or by some of their key operators (such as super- and hypermarkets in organic farming, fair trade, local products, etc.). This can be achieved more or less voluntarily by actors in these alternative systems, who may adopt different approaches: some of them advocate total separation, while others express a desire to put pressure on the 'dominant' system (i.e. the agri-industrial system) so that it will introduce more equality into its trading (e.g. fair trade) or to take greater account of the collective aspirations expressed by these diverse and scattered initiatives.

In the same way, the 'professionalisation' and change of scale of so-called alternative systems introduce problems that are similar to those experienced by the dominant system (greater competition between producers, efforts to achieve better economic efficiency through cost reduction, etc.). Conflicts between alternative systems also lead to strategic alliances that may broaden their objectives and consolidate their positions (the introduction of environmental concerns into fair trade, of the principles of equity in organic farming or short chain systems, etc.).

These systems (both 'alternative' and 'dominant') should be viewed in a context of *ongoing interactions, recombination and hybridisation*. The scientific studies that have analysed shifts of the food industry towards sustainability, in both Northern and Southern countries, ultimately suggest the existence of an innovation model based on the *co-evolution of a multiplicity of food models*.

Innovation processes that place greater emphasis on sustainability within the 'dominant' model thus result from:

- endogenous dynamics (response to pressure/encouragement from consumers and stakeholders and evolutions in the corpus of standards – imposed by public bodies and/or self-generated by major private stakeholders)

- and a permanent comparison with 'alternative food systems' in a democratic and media-driven society.

Support for the emergence, development and adaptation of 'alternative' food systems thus seems to be one of the drivers of food sustainability, alongside actions that could be initiated to influence consumer behaviour, or public and private standards that are a constraint upon the production/processing/distribution of food products.

This process can be seen in an initial approach based on:

- maintaining older systems, subject to certain revisions;
- social movements combining 'committed consumers' and producer initiatives (e.g. organic farming);
- amplification by the media which could lead to institutionalisation;
- integration of criticisms and the appropriation of new systems by major retail distributors and food groups.

The role of the national governments in these processes seems to be markedly contrasted at a global level and according to the alternative systems considered. For example, governments intervene to a greater or lesser extent in defining labels, and sometimes even leave this role to private operators (which is the case for organic farming in many countries, as well as fair trade; see also the problems encountered when trying to define a 'short chain charter', etc.).

4.4   Questions for research

In light of the data presented in this chapter, the principal question for research is 'How can we comprehend and analyse the sustainability of food systems that are both nested in various ecosystems and themselves composite?'. The difficulty resides in the dual vision required: both anthropocentric and eco-centred.

4.4.1   Nesting in complex ecosystems

In order to understand the nesting of food systems in ecosystems, the following approaches are recommended:

- Formal modelling of the dynamics of the three systems: food, energy and chemistry, with variables that include flows of energy, renewable carbon and nitrogen. The challenge is to be able to describe the circular economics of renewable carbon and nitrogen, and to a lesser degree of phosphorus, with spatial and temporal differentiations. The quantification of uncertainty poses the problem of implementing uncertainties in technological blocks, as well as their propagation in the final evaluation process. This issue is essential in order to facilitate decision-making and to identify the most critical obstacles in a process.

- Combination of these representations with balanced economic models.
- Development of environmental evaluation by means of observatories and sample conservation (ecological databank, on soil conservation, for example), geographically associating the different disciplines – including socioeconomic or even societal aspects – and involving future users of the data in the design of systems and the choice of variables to be measured.

For these first two approaches, the principal need is for theoretical bases and numerical tools that can integrate biological/physical/chemical/ecological/ economic processes in indicators and models, covering all scales of time, space and organisation, with the associated laws governing changes of scale. The representation of a human society, and hence its needs and dynamics, will then enable the identification of rules for the optimum viability of natural resources and the optimisation of ecosystem services. This approach must be initiated at a local level, where knowledge is complete and where the global scale is only a boundary condition.

The scale of the biome provides a study context that will enable the consideration of all pertinent mechanisms without being caught up in unmanageable complexity.

4.4.2   Diversity of food systems, comparisons between different models

Through which processes can 'alternative' food systems (traditional or innovative) be maintained or emerge, in both Northern and Southern countries, and contribute to sustainable development?

- What are the (comparative) contributions of different food systems to the challenges of sustainable development? How can knowledge generated on their impacts allow them to evolve?
- How can the co-existence of different food systems enable a better response to the challenges of sustainability, and particularly that of food security?
- Which reasons/claims/social movements will enable the development of alternative systems (social/environmental/nutritional challenges, etc.)?
- What is the role of public stakeholders in the maintenance and evolution of these diverse food systems (standardisation, consideration of this diversity in research, teaching, agricultural and food policies, etc.)?

Do interactions between 'alternative' food systems and the dominant agri-industrial system have an impact on sustainability, in both Northern and Southern countries?

- What is the role of alternative food systems in the innovation trajectory of food companies, and players in major retail distribution and collective catering?
- What forms can (or could) interactions take between alternative systems and the dominant model: incentives, integration, complementarities, hybridisation (with a revision of objectives and player logics), dilution, etc.? What impact do these interactions have on the sustainability of the dominant system?
- Will the integration or hybridisation of alternative systems in the agri-industrial system lead to a reconsideration of their impacts on sustainable development, notably in terms of participation in the governance and integration of its different dimensions?

4.4.3   *At the crossroads between systemic and socioeconomic approaches*

- We need to understand societal demands and mechanisms underlying the development of opinion and changes to behaviour in the field of environmental risks (in the broadest sense) as a result of technical or organisational innovations.

5

Industrial organisation and sustainability

Authors: LOUIS-GEORGES SOLER, VINCENT RÉQUILLART
AND GILLES TRYSTRAM

Contributors: JOËL ABECASSIS, ARMELLE CHAMPENOIS,
VINCENT HOVELAQUE, DOMINIQUE PALLET AND
CATHERINE RENARD

A large proportion of foods are produced today by industrial chains that are confronted by numerous challenges. Like all industries, they need to integrate new requirements relative to the environment and sustainability into their activities. Yet they are also subject to major trends, affecting both final consumption and agricultural production. Indeed, on the one hand, consumers (or at least some of them) are expressing new demands relative to food and how it is produced. On the other hand, current trends at the agricultural level may suggest possible changes in the availability, supply regularity and prices of agricultural raw materials. Finally, the competition in which companies are involved at national, European or international levels raises questions as to evolution in their medium and longer term competitiveness.

The aim of this chapter is to review some of the important features of past evolutions and of the technological and economic 'model' upon which they were based. It is thus 'agri-industrial systems', as defined in Chapter 4, which are addressed here. We will then try to identify, in terms of sustainability, the principal challenges faced by these industrial chains, and finally to clarify a certain number of important themes that need to be investigated in future research programmes in order to enlighten public and private decision-making in this field.

Food System Sustainability: Insights from duALIne, eds. Catherine Esnouf, Marie Russel and Nicolas Bricas. Published by Cambridge University Press. © Cambridge University Press 2013.

5.1   Some important features of changes in the industrial food system

Historically, the prime objective of the food industry was to conserve and preserve essential nutrients so as to ensure their availability throughout the year. This was based on the development of preservation techniques and then the use of stabilising ingredients. An important turning point was the development of cold chains, which circumvented microbiological constraints and allowed long-distance trade of non-stabilised food products. These preservation techniques favoured the development of abundant supplies, particularly of ultra-fresh and dairy products, and contributed markedly to the deseasonalisation of fruit and vegetables.

In order to ensure the constant and controlled quality of end products, the food industry turned towards a policy of assembly based on two principles.

The evolution of the general scheme on which the food industry was based was therefore rendered possible by dissociating two stages during processing: on the one hand, fractionation to break down the agricultural raw material and extract its elementary components to develop intermediate food products (ingredients, additives and technological aids), and on the other hand, the assembly of elementary components to form a consumable food for the end market. This deconstruction/reformulation couple lies at the heart of current industrial processes and operates as described below.

This dissociation has resulted in the emergence of two types of company:

- manufacturers of intermediate food products, where most research and development focuses on improving 'cracking' processes and ensuring the regularity of targeted properties in final products, despite the variability of raw materials.
- assembly companies, where innovations have mainly aimed at considerably enlarging the range of products offered to consumers through actions at an industrial level.

The fractionation of agricultural raw materials has facilitated their common substitution and, to a certain extent, caused the development of stronger links within agricultural raw material markets.

Reducing the cost of supplies has been achieved through the *standardisation of agricultural raw materials*, one of the effects of which has been to limit the impacts of random events on the quantity and quality of products manufactured for industry. At the same time, the sources of raw materials have become more global, notably broadening the original factory supply areas. Starting from highly variable raw materials, genetics and farming practices have driven change towards a smaller number of less diverse materials, which differ less over time. This greater regularity of raw material characteristics facilitates the optimisation of industrial processes, any residual variability in the characteristics of agricultural products being addressed by technological smoothing (addition of ingredients during the assembly phase). This general trend is based on the fact that it is less expensive to reduce hazards by stand-ardising agricultural products than to manage an industrial process that must overcome a marked variability of its raw materials. Indeed, standardisation has been an important vector for gains in upstream agricultural productivity.

The standardisation of agricultural raw materials also implies their homoge-nisation and a reduction in the variety of supplies. The diversification of finished products has also increased considerably, so that delayed diversification has become a marked characteristic of the food system. This has been accompanied by a *downstream shift of the drivers behind the creation of variety in the supply* of products to consumers. Dissociation of the fractionation and assembly stages has thus made it possible to develop a process in which the construction of product variety occurs at the industrial level, and more specifically at the assembly stage. The economic calculation that governs this process is also based on the fact that it enables a greater potential for product variety than simple changes to raw materials, and that it is less expensive, more flexible and more reactive than constructing and preserving variety in characteristics all along the chain from the raw material to the finished product (because of the cost of the co-existence of products with differentiated upstream characteristics).[1]

The *marked drop in the transport costs of agricultural raw materials*, whether processed or unprocessed, has also played a central role in the trends observed. Agricultural raw material resources, which were initially local (and still remain so in part), have become international at little extra cost. All raw materials have become available at all times, with cold chains enabling their preservation. This evolution has been based on the development of new logis-tical systems (transport, cold chain, platforms, technological and logistical technologies, etc.) and increasing competition between different regions, the exploitation of economies of scale at the processing level and the relocation of production (see Chapter 6 for more details on this point).

Finally, *major changes to the distribution of food products* have occurred alongside the transformation of the industrial organisation of different chains. The distribution of food products has shifted and become more concentrated,

---

[1] We only describe here the principal processing trends. However, the example of PDO products shows that other paths have been followed and produced a different response. In the case of wines, for instance, consumers adjust very well to a marked variability over time in finished products.

from small specialised traders to supermarkets that are usually non-specialised. In France, this transformation of the food distribution landscape has occurred within just 40 years: while supermarkets only accounted for 5% of food expenditure in 1970, they now have a 70% market share. The concentration of distribution and the development of retailer brands have tipped the balance of power with industry and upstream producers, and induced radical changes to the management and sharing of value within these chains. This has led to a situation in which downstream players drive a considerable part of the supply chain. Changes to the rules on responsibility in the event of safety incidents have also encouraged distributors to develop their own specifications.

5.2   Consequences of the progress towards sustainability

The trends briefly described above have thus resulted in several changes: (1) product innovation, which lies at the heart of competition, (2) a considerable increase in the number and variety of products offered to consumers, and (3) in relative terms, price evolution in favour of processed foods. Overall, a reduction in the relative cost of food (which is reflected by the smaller share of household budgets allocated to food), improvements in food safety, increases in the variety of products, and support for changes to dietary practices and lifestyles through the availability to consumers of processed (often ready-to-eat) products, are the positive outcomes of this evolution.

Furthermore, an increase in the economic efficiency of the chain has enabled the maintenance of a positive trade balance over a long period, and an industrial body that accounts for nearly 15% of all industrial jobs, and 16% of industrial turnover, in France. However, the trends described above have also had major consequences for the distribution of value within chains, because of increased competition in larger geographical areas between upstream producers and changes to the structure of the chain (concentration of distribution and part of industry). This has been reinforced by recent changes to agricultural policies, which have replaced a policy of guaranteed prices with a policy of direct income support, allowing greater 'freedom' for price determination.

However, as well as the positive effects enabled by modernisation of the chain and the changes that have occurred in recent decades, this agri-industrial development model is now being called into question in two respects.

5.2.1   *An industrial model struggling to achieve further gains in productivity?*

Development of the food industry has been accompanied by increased demands for functionalities which has gradually led to the integration of additional requirements:

- food safety control, the two cornerstones being stabilisation and security, which have provided the foundations for appropriate regulations;
- the control of organoleptic attributes and the search for novel and innovative attributes: the sensory development of products, organoleptic improvements, research on innovative structures and textures, new formulations, etc.;
- the search for nutritional attributes or even health effects, ranging from fundamental nutritional benefits (reduction of salt, sugar or fats, addition of fibre) to research on diet with proven health benefits;
- the concept of sustainability and the energy crisis have introduced a new approach to energy efficiency and the importance of by-products and waste as resources that can be upgraded (including water).

The design (or re-design) of food processing methods from a sustainable standpoint implies a need to take global account of all the constraints and functionalities that are, or will be, imposed. Indeed, historically, changes to these processes have been achieved through the addition of constraints that have gradually reduced the degrees of freedom, to a point at which it appears difficult today to add a further set of criteria without reducing previous constraints. Gradually, opportunities for degrees of freedom are diminishing, either because the level of knowledge remains incomplete, or because the increments of successive functionalities have indeed exhausted the potential for compromises that can be achieved by these operations. Additives and ingredients have only partly compensated for the resulting loss of functionalities.

What are the economic consequences of this gradual integration of demands for further functionalities, particularly in terms of productivity? It is difficult to provide a clear answer to this question at present; nevertheless, the food chain is now faced with obstacles resulting from the relative exhaustion of gains in productivity and a certain deadlock in terms of innovation. Studies on the evolution of competitiveness in the food industry have converged on this point (Bontemps *et al.*, 2013; Chantrel and Lecocq, 2009; Thomas, 2005).

The main conclusions are as follows:

- First of all, an analysis of international trade has revealed a certain erosion of the market share of French companies during the past 15 years. The profitability of the French food industry remains average among the main developed countries, but recent developments have indicated a relative reduction in the competitiveness of this industry, including in some markets for differentiated products (for example, the wine industry). Studies have noted the small gains in productivity achieved by the food industry in recent years. Between 1978 and 2005, while gains in agricultural productivity reached about 2% a year, they were only 0.2% at the level of food companies (Butault, 2008). During a more recent period, 1996–2006, the overall yield of production factors even fell by an average of 0.4% per year. This trend should nonetheless be modulated depending on the type of company; SME have recorded a reduction in their performance, but this was partially compensated for by the good performance of a small number of very large corporations. Productivity has remained almost stable in sectors such as dairy or beverages, but has fallen by 0.7% a year in the meat sector (Bontemps *et al.*, 2013).

- The vulnerability of the chain can also be seen insofar as part of the food industry is built on a 'volume' strategy based on efforts to achieve economies of scale, which is strongly dependent on the availability of inexpensive raw materials. An improvement in competitiveness can be sought through innovation or through the development of technological foods where the micro-structure constitutes the principal vector. Numerous authors have indeed maintained that the micro-structure of foods induces two consequences that are factors for innovation: differentiated sensory perceptions and improved bioaccessibility properties in molecules. However, this supposes major expenditure on research and development, which is not, or not easily, accessible for SME (including the need to qualify the food safety of new products).

- Another point concerns employment. During recent decades, the industrial system has increasingly provided and then maintained a large number of jobs in France (currently around 450 000). Two explanations can be given for this trend in employment. Over the long term, jobs that were previously 'concealed', because they corresponded in the past to activities carried out at home, have become visible because industry has now taken over some of these tasks (thanks to innovation and the development of new products, etc.). In the future, the question will arise of the balance between what is done at home versus what is done in the factory. A second element is linked to the levels of capital/labour substitution, which are probably lower than in other industries. One of the challenges in terms of sustainability also concerns the ability of this chain to maintain employment at the present level.

5.2.2   *An industrial model that should participate in the development of new consumption modes?*

On the demand side, and linked to the general evolution of lifestyles, changes have been seen in the choices made by households in terms of the allocation of their budgets and time, the effects being: (1) growing external-isation of the food preparation function, and (2) growing demand for prepared foods, to which the changes made at the industrial level (and described above) have responded.

These dynamics have not been without consequences for consumers. Some consumer skills previously associated with the choice, preparation and storage of foods, have now gradually shifted to the processing and distribution chain (Dubuisson-Quellier, 2008). Meanwhile, consumers have developed new skills in order to cope with increasingly diversified food supplies, to choose between an increasing number of products and to handle shelf lives that are often very short (reading of labels, knowledge of brands and quality labels, etc.). Furthermore, they have been led to delegate product quality and safety controls to market systems, usually guaranteed by public regulators (Dubuisson-Quellier, 2010).

This externalisation of the food production function from the domestic framework to that of processing and distribution has given rise to a growing distance between final consumers and agricultural raw materials (and their producers) and, more generally, 'raw' foods. This increased distance between the final consumer and foods is spatial, temporal and technological. Even if in the very great majority of countries, most agricultural production continues to be processed in the same country (70% in France), changes to the industrial organisation of different chains has led to: (1) greater complexity of production chains, from upstream to end consumers, and (2) a disconnection between the regions of production and consumption. In addition, dissociation of the fraction-ation and assembly functions has opened the way to changes that have further lengthened food chains and promoted a more technological dimension of food. This is reflected in particular by questions on the use of additives and techno-logical aids. Finally, the time shift between the production and consumption phases, which has been central to its objectives ever since the emergence of the food industry (storage and preservation of foods, notably thanks to control of the cold chain), and the permanent availability of raw materials, have further reinforced the delayed consumption approach, the nature of which has changed in a context of more global trade.

If we add to this the fears aroused by the safety crises of the 1990s, recent debate on the contribution of processed products to the nutritional status of different populations, and then, even more recently, the questions raised on the environmental impacts of food production and distribution methods, we can see the foundations for a challenge to the agri-industrial model from part of the population. Basically, the industrialisation of food has, on the one hand, occurred alongside changes to lifestyle and responded to growing demands for prepared foods and an externalisation of the meal preparation function. Yet on the other hand it has generated feelings of distrust, linked to this loss of control over food production. The greater distance between consumers and foods is an additional factor for a lack of trust in the food system and, at least in part, has driven new food consumption behaviours (organically farmed products, purchasing through short supply chains, local products, etc.), even if their emergence remains limited at present (Dubuisson-Quellier, 2009a, b).

How can the questions raised by a fraction of consumers on the conditions for food production be taken into account? What are the actions that need to be targeted, by both public authorities and private operators? In practice, the most important trends depend on the interpretation of the changes described briefly above. It is possible that these new food consumption behaviours only reflect the demands of a particular section of the population who are more willing to pay for additional guarantees in terms of health and the environment. Under this hypothesis, the answer lies mainly, if not exclusively, in a segmentation of the market, in order to offer the consumers who wish so, i.e. who are prepared to pay for them, the products they seek (on condition, of course, of a satisfactory guaranteed minimum at both the health and environmental levels for all food supplies). This option mainly raises the question of the identity of the products sought (collective and private brands) and the appropriate systems for guarantees, control, and sanctions relative to this identification/non-identification (see Chapter 10 on the current lack of methods and indicators). It opens the way to the creation of two, relatively distinct markets, thus raising the problem of unequal access to good quality food and thus issues of inequality and equity (and also sustainability at the social level). It could also be considered that these new food consumption behaviours – although they may be quantitatively marginal at present – nonetheless reflect broader expectations, including from those sections of the population who cannot translate them into purchases because of budgetary constraints. In this case, the challenge is more fundamental, as it confronts the agri-industrial food system with the need to restore the link between food and consumers, between agricultural products upstream and finished food products downstream.

We have limited this section to consumer demands and actions. The consequences of the constraints fixed by society (transmitted/anticipated by governments) must also be analysed: reductions in GHG emissions, pressure on resources and the environment, biodiversity, etc.

5.3   Questions for research

As some authors have already pointed out (Lowe et al., 2008; Phillipson and Lowe, 2008), the questions raised today regarding changes to food systems require further research at the interfaces between the food sciences (including nutrition) and processing methods on the one hand, and the social sciences on the other. At least two areas merit further consideration in this respect:

- The understanding of consumer behaviours, and particularly how foods are used at a domestic level, and how households manage waste and products at the end of their life, which in return could help to drive research in terms of formulation and processing methods or the packaging of products.
- The links between industrial choices and productivity might also justify investigations at the crossroads between economics and technologies. The principal options adopted previously (deconstruction/reconstruction of functions within chains (fractionation-assembly, etc.), upstream standardisation and 'delayed differentiation') were all based on economic choices whose robustness in the future needs to be reassessed.

As well as these major orientations, several areas need to be studied in more detail in the fields of technology and economy, as explained in the two sections below.

5.3.1   Process design, technological pathways, industrial organisation

We have mentioned above that the evolution of processes through the introduction of successive criteria has reduced degrees of freedom for industry. Could sustainability constraints be satisfied without having to go back on some of the constraints previously integrated? Is it possible to respond by optimising existing technologies, or is it necessary to more fundamentally re-design food production processes and the organisation of food chains?

Numerous studies have focused on the search for technological alternatives. It is clear that only a few of these studies have led to industrial applications, and that most novel processes, and notably non-thermal decontamination, have remained at a laboratory scale. In fact, there are two principal options that need to be studied in greater detail through new research programmes.

The first option could be to optimise existing technologies in terms of their energy costs, water consumption and robustness regarding raw material use. One interesting possibility, already proposed in the literature, is to reduce the number of state changes that occur during a process. Indeed, not only does the use of steam generate costs and consume a great deal of energy, but many products are dried, rehydrated and then dried again. The idea of a water-free production facility remains a challenge for the food industry.

A second option is to entirely re-design production processes, with two aims: to exploit and generate the underlying functionalities of raw materials without passing through a fractionation state, and to review fractionation so as to exploit fractions that are currently not retained. In particular, a global vision of the sustainability of processing highlights the role of the technological treatment of co-products as contributing to the generation of applications as fossil fuel alternatives: the development of materials, biomolecules and energy (see Chapter 4 for more details).

Are there technological processes that could be applied without fractionation, enabling the use of a raw material in its initial state? Is it possible to functionalise a raw material without fractionation, thus avoiding the use of ingredients and additives? For consumers, will this imply having to accept greater variability in the finished product? Future research needs to clarify these issues.

One interesting option, with a high potential for an impact on sustainability, is to accept that the industrial chain cannot do everything. During the 1990s, consumer demands meant that all efforts focused on practicality and rapidity of use. This resulted in highly finalised products, which required only a short time before use. A product that is not finished at the industrial level, but whose processing is pursued in a packaged form (which already exists for some cured meats, wine and cheese), requiring an appropriate final use that notably involves an ad hoc domestic technology, might be an interesting alternative (there is now a major trend towards new products which combine a food and a domestic technology), and in particular would enable energy savings.

Consumers want foods that retain their natural sensory attributes (colour, texture, flavour) without the use of additives or preservatives. In response to this demand, the food industry has developed the concept of minimal processing, which seeks to limit the impact of processes on nutritional and sensory quality and to preserve the properties of foods without using synthetic additives.

Originally, the minimal processing concept mainly concerned food preservation techniques. Indeed, traditional thermal treatments can be beneficial to preservation or the production of certain aromas, but they may also have undesirable effects on other sensory properties or on the nutritional value of foods. These undesirable effects can be minimised through better control of the technologies applied (HTST[2] treatment, aseptic processes) or the use of novel technologies such as high pressure, pulsed light or even the introduction of new functions in packaging systems (modified atmosphere, introduction of natural preservatives).

Today, the minimal processing concept is starting to spread to all phases of food development. The aim is now to reduce the amount of energy used to produce a food, and limit its processing costs. Indeed, since Mainguy (1989), food has been

---

[2] HTST: high temperature, short time.

considered as a complex system with numerous functionalities. Each of these functions has been gradually developed and refined so that consumers can be offered foods with increasingly specific functions, although these may sometimes mean that they are far from authentic. Each incremental addition of a new function (via the introduction of ingredients, processes or their coupling) also has an impact on the price of a food. If this is accepted, a food is still usually considered as a sum of chemical ingredients, and its production is based more on technological innovation (e.g. functional ingredients) than on a managed scientific approach.

Under the minimal processing concept, the aim is not so much to promote an 'old-fashioned' vision but rather to identify research issues linked to these new consumer demands. By considering that a biological material is by nature made up of molecular assemblies with specific properties, minimal processing then aims to obtain the desired functional properties in a food, based on structures and properties that already existed in the raw materials. In-depth knowledge of the different levels of organisation of the material, and their evolution, over and above their composition, is therefore a key factor in the development of new production processes that will allow the preservation of certain native structures or the development of these usage functions. Knowledge of these structures and their associated properties, as well as their evolution according to mass and energy transfer rules, should make it possible to identify new technological processes that combine minimal processing and controlled processing costs.

Finally, one major question concerns the interface between upstream agriculture and the industrial chain, where two main points can be highlighted:

- Faced with foreseeable changes in the food raw materials market, and particularly a risk of increased variability in their characteristics, manufacturing units will have to deal with an accentuated 'bow tie' effect, which consists in having to manage a greater variety of inputs while at the same time generating varied functionalities for the end product in a controlled manner. Indeed, widely accepted and consensual trends highlight the search for tailor-made products adapted to specific classes of consumers and associated with functionalities that are themselves specific. The role of the process (including formulation) becomes crucial under this vision, in which the challenges of robustness and flexibility are of considerable importance.
- The standardisation of agricultural production and the use of industrial drivers to create variety among end products raise the above-mentioned problems. To what extent will upstream agriculture be able to contribute once again to creating variety among products? Which 'differentiating' characteristics can act at the level of agricultural

supply, and what contribution can downstream links make through the additional functionalities constructed by processes?

In any case, these questions offer a new approach to the relationships between agriculture and industry.

5.3.2   *Economic efficiency of companies and chains and the constraints of sustainability*

One of the more general questions posed is that of changes to the economic model – outlined above – in the face of growing demands for sustainability, whether these changes result from public regulations or from the expectations of consumers and other pressure groups. Under which conditions will companies and chains be able to integrate all these demands, and at what cost?

In general, two views are opposed regarding this question. Some consider that a reinforcement of environmental constraints may induce a rise in costs that will limit gains in productivity and the economic efficiency of companies. For others, the objective of reducing environmental impacts may create incentives to innovate and induce productivity gains that will, at least to some extent, compensate for the additional costs of environmental adaptation. In either case, the overall situation, i.e. comprising environmental progress, must be assessed.

Gains in productivity can first of all result from a reduction in raw material prices, a reduction in energy consumption or a reduction in capital and labour costs. Such effects have been demonstrated in other industries, but to our knowledge, the possible impacts of these cost reductions on the efficiency of the French food industry still need to be evaluated.

The other option is to take account of environmental investments in downstream chains: making such investments could become a condition of access to certain markets or allow the development of differentiation strategies to target consumers. On this latter point, evaluations to date have led to the adoption of a measured position. So far, a relatively small fraction of the population has been willing to pay higher prices for more environmentally friendly products (see Chapter 2 for more details). In some cases, it has been seen that emphasis on the environmental dimension may have reduced willingness to pay for conventional products rather than increasing willingness to pay for more environmentally friendly products. In this context, the development of environmental labelling on food products could have a significant impact on the industrial organisation of the chain, less by modifying consumer choice than by making the environmental dimension a driver for industrial competition.

Adaptation to these new constraints raises the question of changes to company competitiveness at the national and international levels and, based on the implications for the development of these chains, its impact on employment (the food industry is one of the two largest employers in Europe). One important research area concerns the impacts of different sustainability constraints on the productivity and competitiveness of individual companies and chains. As well as measuring the effects of sustainability constraints on productivity, this also concerns issues of competitiveness between companies exposed to these constraints at different levels, particularly regarding the implications for employment.

The distribution of value within chains is also a major challenge, particularly since upstream prices will probably vary more markedly than in the past. This raises questions regarding risk sharing, price transmission within food chains and the long-term future of the weakest stakeholders. More generally, we need to better understand the development and dynamics of price structures within chains and how the resulting value is shared. In particular, better characterisation of the following is necessary:

- Transmission of price shocks in a context in which greater variability in agricultural product prices is anticipated.
- The impact of the strategies adopted by industry and distributors in terms of establishing private standards for their relationships with upstream suppliers (exclusivity, increase in negotiating power, etc.) and the interactions between these private choices and public policies on standardisation.
- The strategies of food and distribution companies regarding their access to customers (role of brands, possibility of direct customer access for food companies, etc.) and more generally an analysis of both cooperation and competitive relationships between food companies and distributors. It is in both these groups' best interest to ensure that the global market is as large as possible (cooperation) but also to gain the largest possible market share (distribution of value). The emergence of new distribution channels for food products should also be considered in this context, whether in the case of short chains, e-commerce, or the development of retail outlets by food firms themselves.
- More generally, public sector demands relative to sustainability seem to be pushing firms to move in this direction and highlighting their social responsibility. Under which conditions, and with what effects?
- Finally, it is necessary to evaluate public policies (both *ex ante* and *ex post*), whether they target supply (standards, regulations) or demand (information, labelling, consumer tax policies), and then to clarify their ability to support, encourage or limit (if necessary)

the changes required in terms of the sustainability of food consumption and production. Globally, there has not yet been a true food policy that takes account of sustainability challenges, including the important role of food in health. The situation at present is more a juxtaposition of policies (agricultural, environmental, etc.), which sometimes lack consistency. It is therefore necessary to provide the means to define the most appropriate policy mix that will ensure food sustainability.

5.3.3   Summary of questions for research

Can sustainability constraints be satisfied without having to revise some of the previously integrated constraints? Is it possible to respond by optimising existing technologies, or is it necessary to ensure a more fundamental re-design of food processes and a reorganisation of food chains?

Are there technological processes that can bypass fractionation and use raw materials in their original state? Is it possible to functionalise a raw material without fractionation, so as to circumvent the need for ingredients and additives? Will this mean that consumers must accept greater variability in final products? Faced with predictable trends in the raw materials market, and particularly the risk of increased variability in their characteristics, how can food companies handle an accentuated 'bow-tie' effect, which means they will have to deal with greater variability of raw materials while generating varied functionalities for the end product in a controlled manner? Understanding how a process (including formulation) can contribute to meeting the challenges of robustness and flexibility has become a major question for research.

To what extent can agriculture make a new contribution to creating product variety? Which 'differentiating' characteristics come into play at the supply level, and what can downstream actors contribute through additional functionalities constructed by processes? How will an adaptation to new constraints affect company competitiveness and employment at both the national and international levels? How might this adaptation also affect the sharing of value and risks, the transmission of price shocks within chains and the long-term future of the weakest actors?

What effects can (or could) public policies have, whether they target supply (standards, regulations) or demand (information, labelling, consumer tax policies)? How can the coherence of agricultural, nutritional and environmental policies be ensured, and which policy mix might be the most appropriate in the context of sustainable food policies?

6

Urbanisation and the sustainability of food systems

Author: CARL GAIGNÉ

Contributors: DANIÈLE CAPT, ELISABETH FAGUER, LISE FRAPPIER, MOHAMED HILAL, VINCENT HOVELAQUE, TRISTAN LE COTTY, LAURENT PARROT, BERTRAND SCHMITT AND CHRISTOPHE SOULARD

When dealing with the sustainability of food systems, the spatial dimension merits particular attention. On the one hand, growing urbanisation and urban sprawl in different parts of the world raise questions regarding the sustainability of food supply systems for urban populations. On the other hand, the location of different activities in food systems has a strong effect on the environmental assessment of food systems.

This chapter is distinctive from current scientific literature insofar as the sustainability of food systems is analysed, first, in terms of food supplies to cities, and second, by focusing on location strategies relative to production, processing and distribution activities linked to urban dynamics.

6.1 Sustainably feeding large cities: a major challenge

The challenges of food sustainability cannot be fully understood without taking account of the spatial dynamics of consumption, distribution, processing and production activities in different countries of the world. On the one hand, changes to diets and consumption practices have exerted a major impact on the spatial organisation of the agricultural and food sectors and hence on greenhouse gas emissions and the amount of energy consumed for the shipment of commodities. In addition, the trend towards a geographical concentration of food demand, linked to the urbanisation of populations, has markedly modified both the

Food System Sustainability: Insights from duALIne, eds. Catherine Esnouf, Marie Russel and Nicolas Bricas. Published by Cambridge University Press. © Cambridge University Press 2013.

distances travelled by food products and supply modes because of the development of major retail distributors, and has reoriented land use according to its location. The task of providing sufficient food for all the cities in the world is thus a challenge that requires coordinated interactions between the producers, transporters, traders and the innumerable retailers of food products.

Access to healthy food in sufficient quantities has become a major challenge for large agglomerations. At least two reasons can be put forward. First of all, large quantities, requiring vast areas of productive farmland, need to be imported each day to feed the major conurbations. Indeed, about 6000 tonnes of food need to be imported each day into cities of at least 10 million inhabitants such as Tokyo, São Paulo, Mexico or the Île-de-France region around Paris (FAO, 1998). According to our calculations, about 0.18 hectares of crops and 0.12 hectares of permanent grassland are necessary to feed one French person. In other words, 3 million hectares of agricultural land are required to feed the 11 million people in the Paris region, or six times the agricultural land available locally. Furthermore, by promoting social and spatial exclusion, urbanisation may also give rise to 'nutritional' exclusion, in other words, insufficient access to a satisfactory diet that can achieve a nutritional status that is favourable to health. Indeed, areas with a concentration of low-income households in the major cities of wealthy countries account for a not inconsiderable fraction of the overall population. In developing countries, the proportions may even be higher. It is estimated that about one-third of the global urban population lives in what could be qualified as slums.

The decision to focus on an urban issue can be explained by the increasing urbanisation throughout the world. More than half of the global population now lives in urban areas, according to the most recent United Nations report (United Nations, 2007). Even if a strict definition of a city is used, nearly 40% of the global population lives in a metropolitan area of more than 1 million inhabitants. The number of cities with more than a million inhabitants more than doubled between 1950 and 2007, rising from 181 to 414. Today, about a sixth of global GDP is located in 25 cities (New York and Tokyo accounting for 4% of global GDP). This trend towards urbanisation is set to continue in the decades to come. However, urbanisation can have different characteristics, depending on the continent where it occurs. Although the rate of urbanisation is high and stagnating in wealthy countries, there remains a trend towards a spatial spread of towns to the detriment of agricultural land. According to FAO data, the European Union lost 30 million hectares of agricultural land between 1961 and 2003, which represented a net loss of 770 000 hectares per year. By contrast, although African and Asian countries have a relatively low level of urbanisation (even if China and India are home to about 25% of urban dwellers on the planet), these countries are seeing very strong urbanisation trends. The proportion of the total urban population in Africa and Asia increased 2.7-fold between 1950 and 2007. Over the same period, the proportion of the urban population in Africa increased at an average rate of 4.3% per year. Another example comes from Asia, where Dhaka in Bangladesh, with a population of about 9 million inhabitants, is growing at an annual rate of 5%, which means 1300 additional people each day (according to the United Nations).

Analysis of the link between the sustainability of food systems and major cities, their dynamics and their types must take account of the contextual differences, notably between developed, emerging and developing countries where urban and accessibility issues are not always posed in the same way. For example, although different parts of the world are seeing urban growth, this does not automatically mean a convergence of lifestyles and development. The rapid urbanisation of Africa is associated with purchasing power that remains very low or has even deteriorated in some cities. Consequently, and because of simultaneous growth in the rural and urban populations in many African countries, urbanisation in Africa can be accompanied by the development of rural-type lifestyles where urban households without any employment, or with very low wages, turn to agricultural activities to feed themselves or generate income. In contrast, some countries have experienced a simultaneous urbanisation of their economies and a rise in average income, which has driven the emergence of new consumption practices or new food systems.

6.2   Urbanisation, food and sustainability: what are the challenges?

Although the notion of sustainable food systems is recent, its link with urbanisation and the location of the consumption and production of goods and services has not been the subject of many studies in the academic literature. Yet there is considerable data that *identifies the challenges of food sustainability in relation to spatial dynamics*. Work on the location, urbanisation and *transport of goods* has thrown interesting light on the problems linked to energy consumption and greenhouse gas emissions (Cowell and Parkinson, 2003; Pretty *et al.*, 2005). This issue is also addressed in the literature on the link between urbanisation, residential location and *household mobility*. However, this question should not be reduced to just the distance covered by products (food miles) or by households. Other dimensions need to be integrated. Studies of food systems have highlighted other challenges for major cities concerning their relationships with neighbouring regions in a context of *sustainable local development and securing food supplies for urban dwellers*.

6.2.1   Location and transport of food products

Although production activities are the main source of GHG emissions, transport also makes a major contribution. Two examples: as recalled in a study for the ADEME, the transport of the yoghurt consumed in the Paris region (including logistical costs) accounts for one-third of the GHG emissions of this product (Rizet and Keita, 2005). In addition, according to the carbon balance for the City of Paris, the transport of goods to Paris contributes 27% to overall CO_2 emissions. Generally speaking, although agricultural production is a source of many pollutants, the transport of agricultural goods and foodstuffs also generates significant environmental pollution and consumes a great deal of energy.[1] The farming sector, in its broadest sense, is the second largest user of transport. With nearly 57 billion tonne-kilometres (BTKM), flows of agricultural goods and foodstuffs in France accounted for a fifth of road traffic in the country in 2008. In tonnage terms, cereals, on the one hand, and goods grouped under the heading of 'other fresh or frozen vegetables and fresh fruit' on the other, were about equal, each accounting for one-third of the total amount in weight of agricultural goods. In traffic terms, however, fruit and vegetables took the lead by far; they alone accounted for more than 50% of traffic in 2005.

6.2.1.1   An increase in the average distances travelled by goods...

It is generally accepted that demographic growth and changes, together with reductions in the costs of transport, improvements to transport infrastructures and increases in the speed of goods transport during the twentieth century, have all contributed to the growth of the transport of agricultural goods and foodstuffs, as well as of all manufactured products (Gaigné et al., 2011). Another explanation, less well known but far from negligible, resides in the evolution of economic geography that has induced an increase in the average distances travelled since 1975. While the transport of goods in Europe has continued to rise since the 1970s (within 40 years, it has risen from 1409 to 3152 BTKM, or by 124%), the average distance covered by land transport (regardless of type) has also increased by 37.5%, from 72 to 99 km (Sauvant, 2002). A relative stagnation in the average distances travelled has however been observed since 2000. Agricultural and food products are also affected by this increase in the mean distances travelled for long periods. The transport distances of foods and agricultural

---

[1] The transport of goods and people (i.e. the entire transport sector) is a major and increasing source of greenhouse gas (GHG) emissions. More than 20% of all GHG emissions in the EU and USA result from the physical displacement of goods and people. More worrying is the fact that this proportion rose by 28% during the period 1990–2006. Road transport is responsible for 80% of GHG emissions by the transport sector (two-thirds of which are attributable to the use of private vehicles) (EEA, 2007; EPA, 2011; OECD, 2008a).

products have risen by 17% and 32% respectively, since 1975. Similar rises have also been observed in Germany, the Netherlands and the UK.

6.2.1.2 ... and an increase in the mean distances covered by road transport

According to Savin (2000), road transport saw a marked increase in terms of the average distances travelled by goods in many developed countries. Take the example of France: the average distance shifted from 50 km to 85 km between 1975 and 1995, or more than 70%. The transport of goods by rail also increased in terms of the average distance travelled, but to a lesser extent (an increase of 22%, i.e. an approximate average of 350 km at the end of the twentieth century). It should be noted that river transport saw a slight, 5.5% reduction in the mean distance travelled (from 132 km to 125 km). As for food and agricultural products, the average distances travelled by lorry have also risen since 1975, by 33% and 66%, respectively, but at the same time there has been a reduction in the average distance travelled by rail. It should also be noted that regarding the transport of fertilisers, the average distance travelled, using all forms of transport, has fallen by 24%, but their road transport has increased markedly (+73%). In addition, the international transport of agricultural and food products by French carriers almost tripled between 1975 and 1995. The distance travelled internationally by food products has increased little by road (+2%) but has fallen by 27% for rail transport. For agricultural products, increases in these two main transport options have been seen, by +32% and +9%, respectively.

Several reasons can be put forward to explain the increase in the average distances travelled by goods. First of all, changes to food consumption behaviours are one of the main causes of this clear increase. The economic geography has also evolved, because of: (1) changes to the location of activities, thus modifying the specialisation of regions and the spatial dispersion of major companies, and (2) changes to logistical management.

6.2.1.3 ... linked to changes to dietary practices

The demand for agricultural goods has both intensified and diversified. As a result, the volume and lengths of traffic flows have increased considerably. For the same reasons, the growing urbanisation of developing countries has caused a rise in the average distances travelled by agricultural goods. As spatial concentration of the population increases, more food needs to be transported to meet the needs of urban households, thus generating more pollution linked to goods transport. In addition, urbanisation may coincide with a rise in income for a large share of the urban population, as has been observed in many developed countries, which then triggers a rise in food demand and changes to the types of food products required. The emergence of this demand for new products then gives rise to the import of products from increasingly distant regions, generating even more transport. In addition, urbanisation and a rise in income can increase the proportion of processed products consumed by urban households, further encouraging the transport of goods (because of the relatively larger number of goods included in these processed products).

6.2.1.4 . . . and also to regional specialisation

Paul Krugman (1995) showed why, since the industrial revolution, economic development has been unevenly distributed spatially. The grouping of activities and the specialisation of regions resulted from a combination of economies of scale (the more one produces, the lower the costs of each unit produced), reductions in transport costs and the mobility of workers, capital and suppliers. Without denying the importance of the resources available, the gains resulting from a geographical concentration of production becomes a central element to explain regional development. The idea is simple: suppliers are encouraged to be based near their customers so as to reduce their prices, thanks to the lower cost of transporting goods. This reduction triggers higher demand, thus allowing production at a larger scale. Increasing the level of production lowers average costs because of economies of scale. Rising production and falling average production costs enable a reduction in prices. In return, processors are encouraged to move closer geographically to their suppliers. Their geographical proximity enables a reduction in the price of intermediate goods, shorter delivery times for goods or services and the possibility of obtaining the precise characteristics of the desired product more easily. Thus less expensive, more rapid and more efficient transport systems have stimulated trade (Combes and Lafourcade, 2005) and favoured specialisation, which in turn has enabled economies of scale and the specialisation of regions. These mechanisms provide a good explanation, for instance, for the specialisation of Brittany in industries linked to animal production. Certainly, the abundance of labour, and its low cost – arguments often put forward – originally contributed to the development of this livestock sector in Brittany. The important point here concerns the mechanisms that explain the development of Breton specialisation in this livestock sector, because of economies of scale and geographical proximity between firms (Gaigné et al., 2012). As a general rule, many studies have shown that even a slight advantage can generate a grouping of activities, leading to the strong specialisation of a region. Economies of scale and interdependence between firms lie at the heart of this explanation. Other advantages linked to the geographical concentration of activities have also been identified. The spatial proximity of producers belonging to the same sector of activity induces frequent contacts between managers or decision-makers. The transmission of information on markets or suppliers, as well as technical or organisational innovations, are thus facilitated. Frequent contacts also allow suppliers or customers to build up trusting relationships, which are essential when contracts cannot predict all eventualities. Similarly, numerous studies have shown that the geographical proximity of public sector research institutions favours the dissemination of their findings (Fujita and Thisse, 2002). In other words, the efficiency of a company improves with the density of firms in the region where it is situated. Furthermore, it is worth noting that a geographical concentration of firms means they can benefit from a dense employment area. Because of the broad range of qualifications available, an employer who is trying to fill a vacancy will have a greater probability of finding someone with the necessary skills, and reciprocally, workers have a greater chance of finding employment that will exploit their skills. All this results in higher levels of productivity. The existence of a strong local labour market also favours the mobility of workers between firms. By changing jobs, these workers favour the circulation of knowledge between companies based in the same region.

Thus the distances travelled by products have increased not only because of the reduction in transport costs seen during the twentieth century, but also because of gains generated, first, by the economies of scale and geographical proximity, to exploit the economies of agglomeration, and second by the spatial division of labour, to benefit from the comparative advantages of different production areas. Furthermore, a recent study showed that an increase in international competition has induced a spatial reallocation of food production away from regions of consumption, thus increasing the distances travelled by the foodstuffs required to feed large conurbations (Bagoulla et al., 2010).

6.2.1.5   ... and the evolution of industrial strategies

Recent decades have also seen the development of a functional specialisation of regions because of changes to the industrial strategies of processors and distributors. One example is a functional division of labour between urban and rural regions. The strategic functions (research and development, marketing, etc.) of companies are localised in major cities, while production units are based in much less urbanised areas (Fujita and Thisse, 2002). Companies therefore benefit from the advantages of a major city for their strategic functions, without suffering the urban costs that will mainly penalise production-related activities. Furthermore, geographical proximity between production sites and premises for strategic functions facilitates the frequent trips that are necessary for exchanges between decision-making and operational centres. In other words, the flows of goods and employees within companies have probably increased markedly during recent decades.

Since the mid-1990s, the notion has also emerged of global logistics, or a supply chain, which represents all the companies, suppliers, customers and distributors involved in the processes of manufacture, distribution and sale (Melo *et al.*, 2009). Two main principles define this supply chain: the principles of physical flow and information flow. If all industrial sectors are taken together, companies devote an average of 10% of their turnover to logistics (including transport), even if marked disparities remain (an average of 13% in the food industry). In the food sector, the supply chain has evolved considerably in terms of its structural dimensions (storage, warehousing, transport) and strategic dimensions (source of reductions in production costs). These changes are mainly linked to how major distributors are organised.[2] First of all, a distribution network in the form of warehouses and platforms has profoundly modified the geography of food product flows. Second, the use of just-in-time management methods has become widespread, causing more transport of goods. Indeed, when managing flows between stakeholders, strategies to minimise stocks have led to a higher frequency of deliveries between suppliers and customers. Transport has thus become a crucial link in the supply chain. Major research programmes have highlighted the optimisation of loading[3] (impact on unit costs and on the number of lorries necessary[4]). In addition, the management of urban delivery rounds has developed to take account of new transport methods. In rough terms, high volumes of traffic out to the periphery of cities have become the norm, as has the supply of retail outlets by a fleet of smaller vehicles (sometimes electric).

6.2.1.6   An increase in individual mobility within metropolitan areas

If the spatial ratio between production and distribution activities has markedly evolved, the same applies for households. The growing dissociation between places of residence, places of work, places of consumption and places of supply has markedly contributed to an increase in individual mobility. Within a 20-year period, there has been a 20% rise in the average distance travelled by households in all French regions. Although daily trips occupy a constant time budget, their spatial scope depends on the speed of travel. In cities throughout the world, whatever their level of development, the average daily travelling time is approximately 1 hour, with a financial budget of 5% for households without a vehicle and 15% for those with a vehicle. Under these conditions, improvements to the speed of transport do not reduce the duration of

---

[2] More than 75% of food products pass through major distributors.
[3] In 2005, empty lorries accounted for an average of 25% of goods vehicle traffic.
[4] On shop shelves, average inventory shortages due to insufficient lorries reach around 9.5%.

travel, but allow people to travel further. According to the transport and travel surveys for 2007–08, the average distance per trip was about 8 km in 2008 (Hubert, 2009). Individuals make an average of 3.15 local trips each day (within less than 80 km of their homes), and devote 56 minutes of their time to this activity. The travelling times devoted to supplying the household (shopping) represent about 10 minutes, a time which is longer for those without an occupation (about 15 minutes). A car is used for 65% of local trips, which on average represents two daily trips per person. The increase in the average distance travelled by households has tended to occur in small- and medium-sized towns rather than major conurbations, with an increase of more than 26% between 1994 and 2008 in these former towns. In developing countries, the increase in the average distances travelled by households is exponential because of rapid urbanisation and the development of metropolises that is often not, or only poorly, controlled. In many developing countries, the spatial spread of towns is increasing rapidly and the use of cars is growing even faster. Controlling urban development is a major challenge in order to control individual travel flows.

Finally, the location of retail outlets is also an important factor to be taken into account when analysing the spatial organisation of a major city. For example, retail outlets are unevenly spread throughout the Île-de-France region around Paris. Large supermarkets and hypermarkets (with a surface area of more than 300 m^2) covered a total of 7 million m^2 in 2003 (one-third devoted to food), and more than half of them were situated in the outer suburbs (57%). There thus remain persistent regional imbalances in commercial outlets in terms of serving the local community, particularly for vulnerable areas outside a conurbation. A commercial network (notably of food shops) that is well-balanced at the scale of residential areas would be preferable, integrating the objectives of sustainable regional development (location in the urban tissue, with a limit on ribbon development along roads, etc.) and more environmentally friendly, including in terms of rationalising the supply chain (Anderson et al., 2005).

6.2.2   Urbanisation, food systems and public policies

Since the middle of the twentieth century, the food issue has been less and less visible in Northern countries because confidence has been so strong in the food supply guaranteed by the agri-industrial system (Morgan and Sonnino, 2010). However, this has not been the situation in Southern countries, where threats of famine and malnutrition have continued to affect a large part of the population (Collier, 2007; FAO, 2009a, b). The food crisis in 2008, rising cereal prices and concerns over the ability of the planet to feed more than 9 billion people between now and 2050, have changed the stakes. The scope and rapidity of these changes have driven some authors to qualify the current period as one of diet transition, characterised by the uncertain and contested experimentation of a new global food equation (Friedmann, 2009). Public decision-makers in many countries have focused on the issue of food security (a question that lay at the heart of the G20 discussions in 2011). Similarly, countries that are wealthy but dependent on others for their food, such as the Gulf States or Asian countries, have developed food colonisation strategies in poor countries (Blas, 2009; Cotula et al., 2009; von Braun and Meinzen-Dick, 2009).

Urban areas are in the front line because they are now home to the majority of the world's population. In its recent plans, the FAO (FAO, 2010d) included the 'Food for the Cities' action as a high-priority area. The food issue concerns them in terms of the management of rare resources (agricultural land, energy, water) and also because cities are a privileged environment for expression by citizens. For this reason, many large cities throughout the world (such as New York and London) have also discussed this problem (Morgan, 2009). According to Morgan (2010), the mobilisation of cities relative to food questions was triggered initially by debate on the health of their inhabitants, but has now tended to spread to holistic approaches covering all dimensions of the environment and equality. Some global metropolises have developed full-fledged food strategies, such as the city and county of San Francisco which has laid down 11 principles for healthy and sustainable food policies (Newsom, 2009).

6.2.2.1　What is the role of local food suppliers?

The food policies of cities that are now taking form question once again the dynamics of food systems. Although the 'third age' of food systems (which according to Malassis (1996) is characterised by the predominance of an agri-industrial or agro-tertiary food system) has enabled remarkable progress in terms of the price and health safety of products, it has also generated negative externalities which, in the longer term, may threaten the food equilibrium of populations and the ecological footprint of the planet (according to a report by the London Development Agency (2006), more than 40% of the city's carbon footprint can be attributed to its food system). Consideration of the sustainability of food systems requires a review of the current system and of the relationships between cities and the food industry, or the exploration of alternative pathways. These new options will require the study not only of shorter and more diversified production and marketing systems and their links with the local area, but also of governance modes at the regional, national and international levels.

In the same vein, numerous studies have focused on local food systems – similar to the work carried out since the 1980s in Europe and Latin America on the SYAL – *système Agro-alimentaire localisé*, a French concept meaning 'localised food system' (Muchnik *et al.*, 2008; Perrier-Cornet, 2009) – and, more recently (since 2000) on the re-location or re-regionalisation of food systems, an issue that has taken on increasing importance in the scientific literature, public debate and public policies, mainly in high-income countries. However, these studies have not often focused directly on the urban question, but have provided the foundations for debate on the links between local food systems and urbanisation dynamics, and have reviewed current knowledge. Many of these studies have underlined the ambiguity of the notions of 'local' or 'regional' when associated with the concepts of 'alternative' or 'conventional', 'fair' or 'environmental' products, systems or sectors (Brown and Miller, 2008; Donald *et al.*, 2010; Feagan, 2007; Ilbery *et al.*, 2005), their main objective having been to identify and clarify the terms of the debate and controversies relative to these systems, which is a prerequisite to measuring their importance and their evolution (Darby *et al.*, 2008; Hardesty, 2008). Debate has mainly focused on what local or regional products are, and on the characteristics and performance of local systems compared with the dominant, 'conventional' food system.

Thus the first area for debate, that of defining the scope of local (regional) products, is based on the observation that, unlike the notion of organic products, there is no legal or universally accepted definition of a local (regional) product. This is first of all a geographical concept that refers to the geographical distance between the producers and consumers of food products, but views on the boundaries of what is local or regional vary considerably depending on the areas, consumers and products concerned, and whether they are fresh (perishable) or processed. Furthermore, geographical proximity is only one of the components in the definition of local products, as has been shown by numerous studies carried out in different developed countries on the perceptions and reasons that guide consumers in their food consumption choices, their purchasing practices and their willingness to pay for local products. The reasons for consuming local products are very varied (Carpio and Isengildina-Massa, 2009): consumers value the quality of these products (freshness, flavour) which they associate with production and processing methods, and are sensitive to health, environmental and ethical issues (support for local products and for fair trade).

Linked to these supposed characteristics of local products, the second point of debate concerns the distinctiveness of different types of local product systems, contrasting short and long supply chains. The notions of short and long chains do not have a single, shared definition. A first definition of short chains for local products reflects the reduction in the number of intermediaries between producers and consumers, considered as being associated with geographical proximity: these chains include all types of direct sale, or even some types of indirect sale with only one intermediary. The number of farms selling direct to consumers, and the market share of the products thus sold, have increased during the past decade; this development has varied in different countries and regions (Capt, 2008; Capt and Wavresky, 2010; Martinez *et al.*, 2010), but this segment only accounts for a small proportion of agricultural trade. A second definition simply considers the geographical proximity between the production of local goods and where they are consumed; in this case, account is also taken of local product systems that include more than one intermediary between producers and consumers ('labelled' products or not). The issues related to the choice of a specific definition are generally not explained clearly in the studies concerned.

The literature on local production systems emphasises the fact that the sustainability of urban food policies should not be restricted to using a food miles approach but should also include other dimensions (Carlsson-Kanyama *et al.*, 2003; Coley *et al.*, 2009). Although food policies should have a local dimension in terms of maintaining agricultural production potential in urban regions, they must also have a global dimension. Different reasons can be advanced, and we will focus here on three of them. First, for strategic reasons, it is in metropolitan areas' best interest to diversify their sources of supply. By purchasing food products from other regions, cities reduce the risks of short-ages. Second, even though geographical proximity can reduce the environmental problems linked to transport (an impact that is nevertheless controversial), the ultimate ecological balance may be negative. Indeed, environmental pollution linked to production can vary from one geographical area to another for the same product. It may be preferable to import certain products because the production conditions in other regions are more favourable to the environment. The ecological gains are even greater if imports are achieved using less polluting modes of transport. Third, the ethical dimension is important. The acquisition of food products under 'fair trade' conditions is a good illustration of this. Morgan and Sonnino (2010) spoke in this case of cosmopolitan localism, including the ethnic and cultural diversity of cities – capacious, multicultural and inclusive localism – that they compared to defensive localism. Cities can thus have a dual purpose; they comprise a border-area, open to the rural periphery and endowed with regional identity, and a network-area that includes the countries and regions of origin of its urban populations.

The relationships between urbanisation and food systems have also been described as being ambivalent, insofar as urban development can be both a destroyer of agriculture and a driver for its development (Bricas and Seck, 2004). One debate concerns the current and future importance of agriculture within and around cities to world food security. Highlighted by the FAO (FAO, 2009b), the food role of urban fringe agriculture has also been demonstrated by different surveys. For example, Havaligi (Havaligi, 2009), who studied producers and farmers around San Francisco and Bangalore, urban farmers in California, Vancouver, Mumbai, Sydney and Adelaide, and consumers in the USA, Canada, India and the UK, showed that urban agriculture contributed to adaptation to climate change and constituted an efficient tool to ensure the food security of cities. However, few studies have clarified the links between the local and global levels of food systems, although all emphasise the need for multi-layer governance.

6.2.2.2 The case of cities in Southern countries

The dynamics of urbanisation can play an important role in the development and choice of location of agricultural production through food demand (Boserup, 1975) or the labour market (Ruttan, 2002). This is currently the case in countries where feeding rapidly growing cities is dependent on the ability of both local agriculture and international markets to respond to the growth of urban clusters (De Bon *et al.*, 2010). The urbanisation effects that drive local supply (productivity, land availability, rural population) can be seen in some production sectors, notably market gardening, and in cities such as Lagos, Abidjan or Nairobi (Pingali *et al.*, 1987), but this rule is not systematic. When this local domino effect is observed, it may take the form of urban fringe agriculture (Dury *et al.*, 2004; Temple and Moustier, 2004), or even urban farming (for example, in East Africa, between 17% and 36% of people in urban populations produce food or farm livestock; Lee-Smith, 2010), or of production regions whose location is determined by various factors that include urban development (Bricas and Seck, 2004). Among these determinants, transport costs, and more generally 'market tensions' that reflect the strength of the interactions between urban food demand and local agricultural supply, seem to play an important role. The dissemination of technology may promote the development of productivity and agricultural income close to cities (Tauriainen and Young, 1976). The country's institutional climate also seems to be important, via the propensity of producers to invest in food agriculture (Hayami and Ruttan, 1985).

Inversely, the development of cities may also result in the eviction of farmers because of competition for land use. However, the respective contributions of these different factors to the development of different regions, including in West Africa, to their sustainability, to the modalities of this growth of agriculture (extensification/intensification) and to the food security of rapidly growing cities, are not clearly established. It is probably not possible to judge the sustainability of city supply modes in terms of the location of agricultural production, but we can imagine that food crises and possible responses to economic downturns will be influenced by the supply mode.

6.2.2.3   An illustration for Northern countries: the strategy adopted by the
          Île-de-France region

The strategy adopted by the Île-de-France region (Greater Paris) is an interesting illustration of the local food policies that could be implemented. This concern has only recently been taken into account. Since March 2010, agriculture has been one of the areas included in the remit of the Regional Vice-President responsible for the environment, agriculture and energy. Each year, the Regional Council devotes nearly €10 million to supporting agriculture and the food sector. The challenge of food supplies is vital for the entire population of the Île-de-France region, which has 11 million inhabitants and is markedly urbanised, but where 50% of land is used for agriculture. In order to understand how the Île-de-France region is dealing with this issue, it is necessary to refer to its planning scheme (SDRIF), a document that is mandatory for all French regions. The issue of food is not the subject of a particular section of the SDRIF. However, feeding 11 million people forms an integral part of the challenges that must be met to ensure that the Île-de-France region remains sustainable. It is also interesting to recall that the Île-de-France has never been self-sufficient in terms of food: numerous goods have always been imported from elsewhere in France or from other, more distant countries (coffee, cocoa, exotic fruits, spices, etc.). The current passion for the redevelopment of local food-producing agriculture should not suggest that the Île-de-France can now achieve self-sufficiency. On the other hand, preserving areas that can remain self-sufficient, or even developing new ones, and creating and supporting local production sectors, seems wholly appropriate from the standpoints of traceability, diversification, landscape, services, multifunctionality, know-how, local varieties, local employment, etc. This may contribute to the debates on reducing greenhouse gas emissions, production methods and distribution channels.

6.3   Questions for research

The potential for research is considerable as few scientific studies have deliberately targeted the link between food sustainability and spatial dynamics. Despite the supposedly growing consumer interest in local agriculture, and also that of other actors in the food system (retailers, collective catering, etc.), alongside government and regional programmes, research efforts are still too few and struggle to provide an overview of the importance of current trends, of the diversity of food systems, the factors which influence them, their impact in terms of sustainable development and their links to spatial factors. We propose below some research avenues in two major areas.

6.3.1   *Impacts of rising energy prices on the location of different agri-food*
        activities and, in turn, on the sustainability of food systems in a context
        of increasing urbanisation

Although many studies have been performed on the factors determining the location of agri-food activities (Fujita and Thisse, 2002; Melo *et al.*, 2009), the challenges identified in the section above clearly demonstrate the importance of research which both (1) focuses on the factors concerning the location of agricultural and food activities affecting the evolution of consumer behaviours, the supply strategies adopted by distributors and collective catering firms with respect to processors (local, national or international suppliers), and (2) takes account of the different dimensions (environmental, social, economic, health, nutritional) of food sustainability. Little research has explored the interactions between food, location and the environment. It is necessary to study the relative location of production, processing and distribution activities at different spatial scales (local, national, international) and the flows and modes of transport involved, taking account of the impact of energy prices and greenhouse gas emissions. Are these factors favourable or not to the development of local agriculture, or to a supply of local products? Should production and consumption sites be associated or dissociated? Should regions be specialised or diversified? What are the impacts of upgrading waste on the location of food chains? Yet, the environmental impact of the spatial structure of food systems cannot be approached simply by the food miles method. Indeed, the spatial structure of food systems also involves other agricultural production and processing technologies as well as their polluting effects – which may vary according to their location, this promoting or not biodiversity and animal welfare – on water consumption and waste management.

Supply chain management is an important dimension because it is linked to the sustainability of the value chain (sustainable supply chain, green logistics) (Fritz and Schiefer, 2008; Smith, 2008).[5] The roles of the urban distribution structure (and notably its final link, the 'last 5 km') and logistics are particularly important, as they are essential links between agricultural production areas and processing and consumption sites. Recently, major retail groups have tried to reinvest in city outlets (overhaul of the hypermarket policies in the 1980s). In this context, urban logistics is an essential factor that must be taken into account in the environmental dimension by private sector stakeholders. More generally, Low Carbon Strategies will probably modify the structure of goods distribution systems, firm location and the consumer behaviour. The coordination

---

[5] The literature already contains models for the design and planning of logistical networks that integrate functions to evaluate the carbon footprint.

of numerous small processors serving the same distribution centres (including for collective catering) in order to reduce transport flows, is a major challenge. The pooling of logistical assets by stakeholders – who may even be competitors – also raises some important questions. Based on which criteria could such collaborations develop (quantity, modes of transport, storage sites, consideration of greenhouse gas emissions, etc.)? What are the mechanisms that hamper the development of virtuous partnerships? Furthermore, the environmental effects of industrial strategies should also be taken into account. Will the vertical and horizontal integration of companies generate less travel? Indeed, on one hand, vertical integration encourages a rationalisation of transport flows, but, on the other hand, generates an increase in the number of geographical areas covered by supply and retailing (the fixed charges linked to export and import activities can be amortised more easily). Similarly, horizontal integration in the distribution sector may generate fewer retail outlets, inducing more travel by consumers but less travel by goods. The question of the ownership of goods during their transportation (the producer, processor or distributor?) is important because of its implications in terms of sustainability and the management of losses. Processors that are based at distance from distribution outlets are encouraged to stock their goods close to distributors in order to anticipate demand. This desire for reactivity may generate more or less considerable losses, depending on the products concerned. As a general rule, is there antinomy between stocks and losses? How can these two parameters be controlled in the food chain? These questions thus concern systems with short, local chains as much as those with long chains.

It is also necessary to integrate the effects of relocating agricultural production on the intensification processes that affect agricultural land near cities in both Northern and Southern countries. Few data are available on the dynamics of urban agriculture. In Southern countries, the overall view is as follows. While opportunities to exploit additional arable land are small in Asia, the same does not apply in Africa or Latin America. Indeed, 80% of all arable land reserves are found on these two continents. This probably explains why intensification processes, notably in Africa, are not structurally as decisive as in Asia. However, urbanisation disturbs this finding because of an unbalanced distribution of populations between zones that are distant from, or integrated in, market areas; in the latter case, there is strong pressure in favour of agricultural intensification (De Bon et al., 2010). How sustainable is agriculture in this context?

Water resources merit particular attention. Water is essential for both agricultural production and the food industry, as a fundamental element in food. Its quantitative and qualitative availability must be integrated in any analysis of spatial dynamics. If agricultural land is available close to cities, it will tend to be used for market gardening and if food industries also develop (a sector that consumes large quantities of water, even if the processes involved could make major savings), then the need for water will be considerable. The sustainable management of water resources by cities is a major question in a context of securing supplies and ensuring the regional autonomy of production. The principle of proximity (collection, treatment and waste), the protection of water catchment areas and equitable access to water resources, in terms of both quality and quantity, then become major challenges.

6.3.2   *What are the relative advantages/drawbacks (environmental, social, economic,*
         health, losses/waste) of the different spatial organisations of food systems (notably
         short chain versus long chain) required to feed conurbations?

The notion of a food system (sector, supply chain, value chain) and characterisation of the diversity of systems (making a distinction between agri-industrial systems and alternative systems, according to the diversity of the differentiation attributes of products – conventional, geographical indication, labels, 'fair trade', etc. – and the proximity (or distance) between production and consumption, the number of intermediaries, local or non-local products, short chains and long chains, etc.) remain the subjects of scientific debate and controversy. Nevertheless, this is fundamental to understand the subtle changes underway affecting the configurations of stakeholders organised around differentiated products, and to formulate the research questions they pose regarding the different dimensions (environmental, economic, social, health, nutritional) at play in a context of food sustainability (see Chapter 2).

In this chapter, we have highlighted the fact that the challenges of sustainability lead (and will continue to lead) cities to maintain different sources of supply for their food products because of the diversity of consumer preferences and issues of food security, while at the same time seeking to obtain greater supplies of local products. Hence the research questions posed concern more specifically the factors at play (globally and locally) in the spatial evolution of different food systems and the modelling of spatial interactions between these factors, in order to understand their ability and efficiency to feed large cities. From this point of view, the problems of supplying cities, and any analysis of the factors at play, differ according to the countries concerned (industrialised countries versus developing countries; North versus South). In industrialised countries, the objective of obtaining more local supplies raises questions that are common to all large cities, but with specificities that depend on their spatial configurations (population size, density, local and regional diversity of food systems).

The issue of the environmental impact of different food systems has first of all been addressed mainly in terms of the impact of their energy costs, notably with respect to GHG emissions, in order to determine whether systems involving the supply of local products have a positive environmental impact (reducing food miles or not). On the one hand, some studies have compared the energy consumption involved in travel related to the production and distribution chain, although others have extended this to take account of travel by consumers, while on the other hand, indirect energy consumption (storage, etc.) has been estimated. The systems covered by these studies were broadly defined by the geographical proximity between the sites of production and consumption (and not by the number of intermediaries), and were thus diversified in terms of the characteristics of products and organisational methods (direct sale by producers to consumers, other types of sale with one or more intermediaries). These studies showed that as our knowledge stands at present, it is not possible to conclude as to the advantages of obtaining supplies of local (regional) products. However, these findings remain controversial and studies need to be pursued. In view of the high degree of travel by consumers to where they obtain their food supplies, such studies should notably enable an evaluation of different distribution systems: supermarkets/local shops, restaurants and other catering outlets, according to transport infrastructures. However, this work should not be restricted to using the food miles approach, because the technologies of agricultural production are also concerned, such as processing methods and their more or less polluting effects, whether they do or do not promote biodiversity, animal welfare, water consumption and waste management, etc. In this respect, the different food systems hold different and controversial positions with respect to the challenge of sustainability.

Although logistical management is a major challenge in a context of supply that is more oriented towards local products in order to overcome the problems posed by energy prices and GHG emissions, other important factors also influence the spatial evolution of different food systems and the availability of local products: changes in consumer preferences relative to the attributes of food products, changes in the varieties offered by producers and translation of these trends into local interactions between consumer behaviour and producer strategies.

Thus the development of an industrial food system has been accompanied by a shift downstream (industry and distribution) of the drivers behind the creation of variety in the supply of products to consumers (a considerable increase in the choice of product varieties) at the same time as evolutions in lifestyle. However, these changes have also resulted in the emergence of opposition to the agri-industrial model from some consumers. Will this opposition grow and result in a shift upstream of the demand for variety (differentiating characteristics of agricultural supply) (cf. Chapter 5)? Work on the evolution of consumer preferences (concerning the differentiation attributes of products) needs to be pursued so as to comprehend whether some of the behaviours described (e.g. in France, an unmet demand to purchase food products in the context of Community Supported Agriculture) will continue in the longer term. It then will be possible to identify how willing consumers are to pay for these varieties (price level) and, if necessary, to travel to specific sites to obtain them, and to understand the interactions between consumer demand and producer supply. Proximity does not only have a geographical dimension, there is also a cognitive dimension. Re-internalising externalities, in other words consumers themselves having to manage the negative effects of their food systems as far as possible, constitutes an incentive to reduce these negative externalities (Princen, 1997). The effects of the location of externalities on behaviour are poorly understood and merit further study.

On the producer side, the question is raised of the determinants governing their products and how they may change. Previous studies have certainly provided some lessons relative to the challenges of sustainability: the positive impact of this supply segment on agricultural employment and on the local economy (e.g. the multiplier effect of farmers' markets in Oklahoma was estimated at between 1.41 and 1.78 by Henneberry et al. (2009), using an 'import substitution' process); the small market share of this agricultural supply sector and uncertainties concerning its growth, or the major spatial and temporal variations in supplies from producers marketing food products under short chain systems. However, the factors that influence this supply – spatial and temporal variability – are still insufficiently explored. It has been possible to demonstrate two types of effect: the effect of the proximity of urban and peri-urban areas, and the effect of the production system, which results in major differences between the products sold according to the area (Capt, 2008; Capt and Wavresky, 2010). In order to understand the ability of this system to respond to the demands of cities for supplies of local products, it is important to study how this supply is adapting (and is likely to adapt) to changes in consumer preferences regarding differentiation (and particularly attributes linked to the environmental impact of agricultural practices). Studies on the economic conditions under which this supply is possible are only in their infancy and need to be pursued (generation of data on retail prices, changes linked to this activity and the remuneration of producers), as do those on the social dimension of this activity (workload and quality of life, amongst others). On the other hand, the role of social capital and social networks has been the subject of sociological and economic studies which have stated that food sustainability depends more on social interactions between producers and consumers (the concept of embeddedness) than on geographical distance (Martinez et al., 2010; Sonnino, 2007). For example, studies have shown that the links which have developed between consumers and producers through organisations such as Community Supported Agriculture (AMAP in France), or collective points of sale, etc., are likely to modify dietary behaviour, with consumers paying more attention to their health (eating better and taking account of the seasonality of products in their cooking habits, for example). However, this link has not been firmly established (Ver Ploeg et al., 2009). One of the most promising strategies seems to consist in increasing the consumption of these products in schools. This point merits particular attention, notably in terms of the conditions for the adoption of this consumption mode by schools and local government. A final question also needs to be explored; that of the organisational methods, new stakeholders and governance involved in ensuring that this type of system will be able to respond to growing urban demand.

Finally, the role of new stakeholders in the emergence of local food systems requires investigation: that of local government bodies. Their challenge is to secure supplies of food products (in quantity and quality) or to demand a return to local production or certain practices. The question is to determine the extent to which local agriculture, on the one hand, will be incentivised by the will to secure local food supplies, and on the other hand to establish whether these stakeholders can structure (or restructure) local sectors in order to feed local urban populations. It is also necessary to analyse the range and efficiency of the different possible drivers of action (development plans, control of land use, financial incentives, regulations, governance systems, etc.) that will ensure food sustainability for major urban areas.

Furthermore, urban policies aimed at improving access to food also merit particular attention. What is the link between location, dietary practices and the food products offered by distributors? What are the implications in public health terms (policies on transport, housing and urban planning)? Urban logistics are also an important theme for public stakeholders in major cities. A recent study by PIPAME[6] (2009) identified five strategic variables for action: mutualisation, transport fleets, land use logistics, delivery sites and the commercial structure. Underlying these structures are issues concerning infrastructures (roads, rail, ports), negative externalities (noise and atmospheric pollution, saturation/congestion, etc.), access to facilities (urban, peripheral and rural presence) and employability (distance from work to home). The question is to

---

[6] Interministerial Centre for the Forecasting and Anticipation of Economic Change (*Pôle interministériel de prospective et d'anticipation des mutations économiques*).

determine to what extent the introduction of new organisational schemes will allow greater economic and environmental efficiency.

More generally, the regional governance of cities needs to be studied and revisited in the context of food sustainability. Some authors qualify this as the fourth pillar of sustainable development. However, it is necessary for research to study the reasons for the success or failure of actions implemented in these areas. Which stakeholders directly or indirectly affect these regions? Based on what representations of the problems? What diagnoses have been made of the food systems that need to be developed, and with what visions for the future? What strategies for actions and projects result, and what types of cooperation or conflict appear? Also, what are the procedures, languages and tools (databases, indicators, foresight studies, mediation, etc.) that will enable an improvement in the interactions between stakeholders so that sustainable management can advance in food-producing regions? And finally, how will all this lead (or not lead) to a recomposition of regions based on new relationships with food? One problem is that these are transversal and cross-disciplinary research fields, close to action research, so it is difficult to integrate them in predefined frameworks. However, there is a considerable corpus of information upon which this work can be based: development plans, charters, contracts, etc., drawn up by towns, regions, countries or natural parks, etc., in France and in other countries. Some research disciplines and communities are starting to unite their efforts on these questions; regional sciences, for example, or urban and development studies. A scientific community is thus forming at the crossroads of economics, social sciences and regional sciences.

7

Losses and wastage

Authors: BARBARA REDLINGSHÖFER AND ANNIE SOYEUX

Contributors: ARMELLE CHAMPENOIS AND SOPHIE LE PERCHEC

It is necessary to recognise and understand losses and wastage in order to enable their reduction and recovery.

7.1   Introduction

The importance of food losses and wastage in the context of sustainability and the lack of understanding of their extent, the mechanisms at play and the role of different stakeholders are such that we have devoted an entire chapter to this subject of crucial importance to the sustainability of food systems.

In this chapter we will define the losses and wastage that affect products intended directly for human consumption, in both Northern and Southern countries, without specifying the food systems in which they occur. By analysing recovery options of these lost or wasted products, this chapter will fuel the debate referred to in Chapter 4 on the interlocking character of food, energy and chemical systems and on the circular economy of agricultural biomass.

Losses and wastage are not linked to stakeholders' carelessness. In Northern countries, they are the visible result of socioeconomic changes to food systems at a planetary level (globalisation of markets, industrialisation of processing, etc.), and of changes in value systems (leisure time versus food preparation time) (Soyeux, 2010).

The growth of the world's population, the perception of the finite supply of arable land, the affirmation of the right to food and rises in the cost of agricultural products are all drivers that place human food consumption at the forefront of political and social concerns. Whereas these drivers put pressure on agriculture to increase its production levels, reducing post-harvest losses and

Food System Sustainability: Insights from duALIne, eds. Catherine Esnouf, Marie Russel and Nicolas Bricas. Published by Cambridge University Press. © Cambridge University Press 2013.

wastage forms part of the solution. This at last being considered as an econom-ical and ecological means (Gustavsson *et al.*, 2011; Lundqvist *et al.*, 2008; Nellemann *et al.*, 2009) of increasing food availability in the context of the expected growth of the world's population between now and 2050.

The FAO considers that to ensure global food security in 2050, agricultural production will have to rise by 70% (Bruinsma, 2009). Yet primary production is already considerably constrained by biological and agronomical factors, by randomly occurring natural disasters and by consumer demand. Any recovery that is possible at any point in the food chain would contribute to relieving pressure from primary production. As an indication, a British foresight study considers that at a global level, cutting in half losses and wastage throughout the food chain would lead to saving an amount of food products equivalent to approximately 25% of current agricultural production (Foresight, 2011).

Lost or wasted food could have been used by other consumers (potentially with lower purchasing power), recycled for animal feed, used to generate energy (anaerobic digestion or methanisation) or to make compost. All this food has a cost in terms of land use, water and energy consumption and labour. It uses up scarce resources and generates greenhouse gas emissions.[1] Like the food that is consumed, lost and wasted food has utilised the entire value chain of process-ing, packaging, storage, transport and distribution. It has a cost to consumers who do not draw any benefits from it. In addition, waste generates greenhouse gas emissions because of its collection, processing and deposition in landfills. For all these reasons, losses and wastage in the food system are not sustainable.

The search for methods to reduce losses and wastage is an important driver towards the sustainable organisation of food systems throughout the world.

The subject of food losses is not a recent one. Man has extensive experience in preserving and protecting food resources, and has developed strategies against numerous pests that compete for food. This experience has contributed to the survival of the human species. Such efforts remain a daily battle not only for farmers, and particularly for millions of smallholders in Southern countries, but also for supply chains in Northern countries. Despite several projects and an ambitious FAO programme for Southern countries – Prevention of Food Losses – which was set up following the World Food Conference in 1974, food losses remain as great a challenge today as they were in the past.

At the turn of the millennium, this topic took on a new dimension in the modern and affluent societies of Northern countries due to the increasing

[1] According to a British study by WRAP and the WWF, wasted food products use 6% of all water requirements in the UK, and are the source of 3% of UK greenhouse gas emissions, without taking account of emissions from the value chains of food imports (WRAP and WWF, 2011).

awareness of wastage of perfectly edible foodstuffs, the vast amounts of which are only now beginning to be assessed (Foresight, 2011; Gustavsson *et al.*, 2011; Lundqvist *et al.*, 2008; Nellemann *et al.*, 2009; Parfitt *et al.*, 2010).

Today, major food losses do not occur at the same stage in Northern and Southern countries' food systems: whereas in the former most wastage seems to occur at the level of distribution, of in-home meal preparation and out-of-home catering sectors, in Southern countries, most losses occur following harvest, during storage, transport and primary processing (post-harvest losses).

7.2   What is loss, and what is wastage?

Definitions exist for the terms *losses* and *wastage*:

- Loss (FAO, 1981): 'Loss' means any change in the availability, edibility, wholesomeness or quality of the food that prevents it from being consumed by people. It should be noted that under this definition, food losses can be both quantitative and qualitative (related to food safety, nutrition, cleanliness, purity, etc.) (Tyler and Gilman, 1979). In the context of Southern countries, the term of post-harvest losses is employed: i.e. between harvest and distribution.
- Wastage (Lundqvist *et al.*, 2008): 'The deliberate discarding and throwing away of food that is fit for purpose and perfectly good to eat.' The product is then no longer available for human consumption, even if its initial state would have permitted it. In this context, wastage occurs during the final stages of the food chain: distribution, catering sectors, restaurants, food producers and households.

The broad diversity of situations throughout the world in which losses and wastage occur makes it difficult to clearly distinguish between these two phenomena.[2] According to the FAO definition, foodstuffs fed to animals as they are unfit for human consumption are considered as losses, even though they in return supply the human diet with milk, meat, eggs and fish. Other authors have associated animal feed with losses and wastage in a more (Stuart, 2009) or less (Lundqvist *et al.*, 2008) explicit manner.

The definition of wastage does not take account of the fact that food is thrown away when it has deteriorated, but could have been used in time if the consumer had had the necessary culinary knowledge and home economics skills (purchasing, storage, use of left-over food, etc.) to prevent its deterioration. This wastage

---

[2] Indeed, in English, the generic term *food waste* covers both losses and wastage without giving any further detail (Foresight, 2011; Parfitt *et al.*, 2010).

mainly occurs without the consumer being aware of it. When food purchases exceed needs (portion sizes, sales promotions, etc.), then throwing food away is wastage, but eating it up is also a form of wastage (Nellemann *et al.*, 2009; Stuart, 2009). In the USA, for example, several studies funded by the USDA[3] in the past 10 years, all aimed at determining the reasons for the obesity epidemic, gave rise to estimations of wastage.

A lack of consensual definitions of the terms losses and wastage can partly explain the marked differences that can be seen between estimates of the wasted volumes.

Based on the assumption that wastage covers anything that could have been consumed by humans but which is not, then the quantity of products qualifying as waste is considerable. However, if account is taken of the destination of derived products, a more interesting systemic typology results. Most agricultural raw materials are cracked or refined into edible products, animal feed, products for chemistry, energy generation and soil amendment, as well as solid waste that is burned and liquid effluents.

During the duALIne project, a widely employed distinction was adopted whereby wastage is attributed to Northern countries and post-harvest losses to Southern countries. However, nothing is known of the situation in emerging countries (China, India or Brazil) or major conurbations in Southern countries. With rapid urbanisation and the opening up of emerging countries to foreign investment, and given the rapidly changing organisation of food systems (diversification of distribution channels, market penetration of major retailers, industrialisation of supply chains and diet changes), are these losses in fact becoming more similar to those seen in industrialised countries? Due to a considerable lack of knowledge, our analysis does not cover losses and wastage in urban distribution systems of Southern countries.

7.3   Origins and location of losses and wastage in food systems

Quantitative and qualitative losses are possible at all levels of the food system, for numerous reasons:

- bacteria, yeasts, moulds, insects and rodents can damage products during harvest, storage, transport or as a result of defective or inappropriate packaging;
- temperature, light, oxygen, moisture, dryness and natural enzymes are destructive factors affecting the quality and shelf-life of products throughout the food chain.

---

[3] http://www.ers.usda.gov/Data/FoodConsumption/FoodGuideIndex.htm.

Alongside these biological and physical factors, the state of equipment, road and storage infrastructures, the economic context, hygiene standards and finally modern lifestyles of households give rise to losses and wastage. Figure 7.1 below is a non-exhaustive illustration of their causes and location at different levels in the food systems of Northern and Southern countries. The grey shading indicates the levels of losses and wastage, from pale grey (low) to dark grey (high), at these different steps.

7.4   Poor quantification of losses and wastage, North and South

Since the scope of duALIne did not cover upstream agricultural activities, we will not consider pre-harvest losses at present. Furthermore, our investigations did not concern marine fisheries that, particularly for trawl fisheries, generate considerable fish discard (Kelleher, 2005).[4]

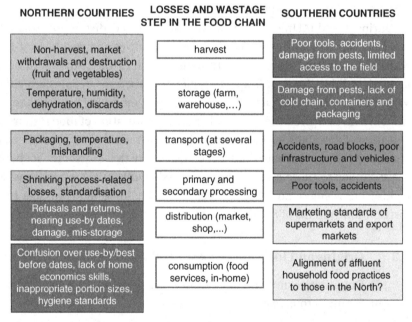

Figure 7.1 Causes and locations of losses and wastage in the food systems of Northern and Southern countries (Source: Illustration by the authors compiled using different sources).

---

[4] The FAO estimates the amount of fish discards at about 7 million tonnes (Kelleher, 2005). However, there is considerable uncertainty over this volume because most of the discard is not declared. In addition, different definitions of fish by-catch and resulting discard exist. Depending on the chosen definition, by-catch may reach 20 million tonnes (FAO and World Bank, 2010).

Several sources estimate the extent of food losses and wastage at a global level to be about 30% of primary production intended for human consumption (Foresight, 2011; Gustavsson *et al.*, 2011; Lundqvist *et al.*, 2008). Based on data in the literature and FAO food balance sheets, broken down by major regions of the world, Gustavsson *et al.* (2011), determine that 208–300 kg per capita per year of food are wasted in European and North American countries, including 95–115 kg wasted by consumers. In sub-Saharan Africa and South and South-East Asian countries, the figures are 120–170 kg per capita per year (including 6–11 kg wasted by consumers).

7.4.1   *In Northern countries*

In recent years, several scientific publications in the USA, UK and Sweden, publicised by consumer associations in Belgium or Switzerland, highlight the issue of food wastage. Eurostat compiled data on the 27 Member Countries in order to gain a clearer idea of food waste produced in Europe, and its evolution over time. These data, reviewed and partially updated in 2009, were used to study waste contributors (Figure 7.2).

According to a study by Jones (2004) in the USA, each household wastes 14% of its food purchases by weight, for an equivalent of $589 per year. Transposed to the 73 million households in the USA, this would represent 17 million tonnes, or be worth $43 billion each year. Catering, fast-food outlets and retailers throw away 27 million tonnes of food each year, with an equivalent value of $68 billion.

The highest waste amounts concern staple foods (and above all bread), followed by vegetables, fruits and meat.

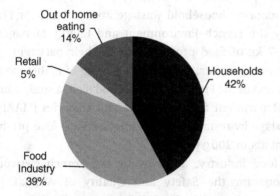

Figure 7.2 Breakdown of the 89 million tonnes of food waste each year in Europe (179 kg/person/year). Source: BIOIS (Bio Intelligence Service, 2010) from Eurostat 2006 and national data for 2006 to 2009. NB. The figure for the food industry also includes by-products.

Table 7.1 *Wastage in the UK in 2009 (avoidable food waste) (Source: WRAP, 2009).*

Type of food	Waste in tonnes
Fresh vegetables and salads	860 000
Drinks	870 000
Fresh fruit	500 000
Bakery	680 000
Meals (home-made and pre-prepared)	660 000
Fish and meat	290 000
Dairy and eggs	530 000
Processed vegetables and salad	210 000
Condiments, sauces, herbs and spices	200 000
Staple foods	200 000
Cakes and desserts	190 000
Confectionary and snacks	67 000
Other, including oil, fats and processed fruits	70 000
Total	5 300 000

In the UK, recent data have shown that 5.3 million tonnes of food and drink are thrown away, two-thirds of which could have been consumed. This wastage mainly concerns fresh fruits and vegetables, drinks, bread and bakery products and prepared dishes (Table 7.1). Its estimated value is £480 per year and per household (up to £680 for households with children), or an average of £50 per month. In total, wastage corresponds to 25% of purchases in volume terms, with some variation depending on the products: 7% of milk is wasted, 36% of bread and more than 50% of salad products.

No similarly detailed data on household wastage are available for France. According to a study by the French Environment and Energy Management Agency (ADEME, 2007), 7 kg of food products still in their packaging were thrown away per person and per year, and a total of 20 kg of edible food was not consumed. In the context of the French National Food Plan, a study commissioned in 2011 by the Department for Agriculture and Fisheries (MAAPRAT) estimates losses and wastage by restaurants, catering services, food producers and the retailing sector at about 200 g per meal.

At the level of the food industry, a survey by the French Network of Professional Organisations for the Safety and Quality of Animal Foods (RESEDA[5]) detailed the processing methods of all food sectors and determined their incurring co-products and by-products. Co-products are used for animal

---

[5] The RESEDA survey took its figures from surveys on different sectors (RESEDA, 2005).

feed, while 'by-products' are used in other industries such as the cosmetics, pharmaceutical and chemical industries, or are used as fertilisers, ballast or energy, etc. The primary processing stage of plant products recovers both incurring co-products and by-products to a considerable extent. The gross volume of these co-products and by-products reaches 9.5 million tonnes of dry matter. Four sectors generate 82% of the gross volumes of co-products and by-products used for animal feed: sugar refinery supplies 33% as distilling dregs and molasses, oil and fat industries provide oilcake (23%), while starch production supplies protein (13%) and milling provides bran (13%).

Meat and dairy industries have greater difficulty in recycling their by-products (0.55 million tonnes) due to their rapid deterioration and to precautionary regulations: two-thirds of this waste, or nearly 0.3 million tonnes of meat and bone meal, have been burnt since the bovine spongiform encephalopathy (BSE) epidemic.[6] The remainder is processed into gelatine and glue.

In France, little is known about wastage at distribution stage, whereas waste arising from restaurants is counted with household waste. According to the *Fédération des Banques alimentaires*, one of France's charities in the field of food donations, wastage of ultra-fresh products and of fruits and vegetables by hypermarkets and supermarkets may be as much as 600 000 tonnes a year, a third of which could be recovered. As for rejected batches, they are the property of the transport company, which sells them at a low price as compensation, or destroys them, when returning them to the factory is not possible or too costly. 'Unpackaging' units are starting to develop which can recycle some of these products for animal feed, but the quantities concerned are unknown.

Donations from retailers to charities reached more than 50 000 tonnes/year in 2010. The retailer Carrefour donated 960 tonnes to associations in 2008, while the *Fédération des entreprises du commerce et de la distribution* (FCD), French federation of retailing and distribution companies, claims that its members supplied 30%, that is around 30 000 tonnes, of the products collected by the *Banques alimentaires* in 2009. Although these donations reduce waste, the quantities are still far below the 200 000 tonnes estimated as recoverable by the *Banques alimentaires* (SITA France and *Banques alimentaires du Bas-Rhin*, 2011).

FareShare, the National UK food donation charity, considers that, if it had the necessary logistical resources, it could recover and distribute 1.6 million tonnes of discarded edible food. Yet, only 2000 tonnes were recovered in 2007, which served to hand out 3.3 million meals.

---

[6] Today, 20 years after the first BSE crisis, and in view of the screening results, the European Commission and Member States are considering the gradual reintroduction of processed animal protein (PAP) into animal feed.

7.4.2   In Southern countries

In Southern countries, and even though the international community became aware of the extent of the phenomenon as early as the 1970s and 1980s, few data are available on post-harvest losses. They were mostly associated with storage and synonymous with 'insect damage' at the time, although they are determined by 'a wide variety of circumstances under which products are cultivated, harvested, preserved, processed and marketed' (Tyler, 1982): these include products and varieties, climatic and meteorological conditions during production and harvest, technical equipment, know-how, cultural and social choices. For example, Schulten (1982) showed that the importance of cereal losses depended to a considerable extent upon the type of plant material (traditional varieties, improved varieties and hybrids). Thus, in the case of maize, losses in weight terms could range from 3% with traditional varieties to more than 20% with hybrid varieties.

The quality of the protocols used for data collection must of course be taken into account when analysing quantitative studies.[7] A single measurement of the losses affecting a product in a particular country cannot be considered as a value to be extrapolated to all products and to all countries, or to be valid for harvests throughout the year. Inexpensive quantification methods must be developed in order to generate representative and easily updated values that will cover all stages of the post-harvest systems in all countries.

For several years now, the FAO and its research and development partners have been investing in the constitution of databases on post-harvest losses in Southern countries. Generally speaking, most loss estimates focus on basic foods, and cereals in particular. Far fewer loss data are available for perishable products such as tubers and roots, fruits and vegetables, though they are more vulnerable to damage, and there is practically no information on animal products such as fish, meat and milk.

A review of current knowledge on losses and wastage was recently published by Parfitt et al. (2010), in which the authors were confronted with the same lack of data we have also experienced. As for the available data sources, most of them date from the 1980s and early 1990s, and only concern rural areas. Generally speaking, the data available in the literature or information systems concern losses of quantity, expressed in terms of weight and/or, more rarely, monetary value.

---

[7] As an example, one error often mentioned is that during storage on farms, the quantities removed for domestic consumption are not taken into account into the estimate of losses at the end of the storage period. Henkes showed that during a 9-month period in Tanzania, losses estimated at 30% fell to 11% when domestic consumption was taken into account (Henkes, 1992).

Box 7.1   Two information networks available on post-harvest operations and losses in Southern countries

APHLIS: The database APHLIS (www.phlosses.net), African Post Harvest Losses Information System, managed by the Joint Research Centre of the European Commission and implemented by the Natural Resources Institute (NRI) and partners, provides information on cereal losses in Southern and Eastern Africa. Within APHLIS, literature on cereal losses in Africa has been reviewed in the light of data quality. A network of local experts of country specific post-harvest systems provides support in adjusting loss estimates to the specific local contexts covered by APHLIS. Hence, APHLIS supplies loss estimates that take account of different post-harvest operations, climatic regions, production volumes and other factors. However, only cereal losses are studied. In the longer term, plans are afoot to extend the database to include other products (e.g. pulses) and to cover other geographical areas (West Africa, Asia, Central America and South America).

INPhO: Another database, INPhO, Information Network on Post-harvest Operations, managed by FAO, CIRAD and GTZ (www.fao.org/inpho), contains a large amount of information on post-harvest operations, including loss estimates. This database covers the post-harvest systems of basic cereals (maize, rice, sorghum) and manioc in Asian, African and Latin American countries. A broad range of technical information on post-harvest operations, the physicochemical and nutritional composition of products and some recipes are available, the aim being to support the development of supply chain activities for tropical products.

Losses of rice, the cereal most widely consumed by humans, have been extensively studied, thus offering an opportunity to compare the extent and origins of losses in the post-harvest systems of different countries. Studies agree as to a global loss rate of around 15% (Grolleaud, 2002; Liang et al., 1993), but variations between countries, climatic areas and processes, without mentioning data quality, may be considerable. During storage, which is often studied specifically, these loss rates could range from less than 1% in a study in Malawi (Singano et al., 2008a, b) to 12–13% in Bangladesh (World Bank, cited by Grolleaud, 2002), including a range of 3–6% in China (IDRC, cited by Grolleaud, 2002) or Malaysia (FAO, 2007). The compilation of studies on post-harvest losses of rice by Parfitt et al. (2010) illustrates this variability in available data.

For sub-Saharan Africa, post-harvest losses of cereals prior to processing have been estimated by APHLIS at 10–20%, for a value of around $4 billion. These losses represent 13.5% of the total value of cereal production in these countries (World Bank, 2011).

For perishable products (roots, tubers, fruit, etc.), even fewer representative data are available. In the case of manioc, losses in traditional systems can reach 45% (Jeon and Halos, 1991), while for yams they can reach as much as 50% (Osunde, 2008). As for fruits, post-harvest losses may be between 15% and 50% in developing countries (Coursey and Booth, 1972; Jeffries and Jeger, 1990; Subrahmanyam, 1986). In the Philippines, as a general rule, they can range from 15% to 35% (between 30% and 60% for fruit such as papaya). These figures are relatively representative of the average post-harvest losses recorded in the region covered by the Association of Southeast Asian Nations (ASEAN).

7.5   Lessons from the past, courses of action and strategic orientations

Even if accurate data on food losses and wastage are lacking for most countries of the world, global estimates of their extent indicate a not inconsiderable potential for reduction. However, only global awareness and behavioural changes among stakeholders will be effective. Drivers and courses of action exist in both Northern and Southern countries, but purposeful public policies and solid research remain necessary to support the efficiency of reduction measures.

7.5.1   Northern countries

Changes are underway in Northern countries that reflect a change in stakeholders' perception of the problem and strategies, as well as a growing awareness of the need to deal with what increasingly seems to be a real public concern.

Agriculture and the food industries have already adopted policies to limit losses for cost reduction. A transfer of the use of co-products and by-products away from animal feed towards the human diet is the subject of applied research. The food industry is rediscovering the principle of 'eco-industrial parks' in which the waste of some becomes raw material for others (see more detail in Chapter 4).

As far as logistics and storage are concerned, significant progress has been achieved with respect to containers, stock management and automated order preparation. Packaging needs to evolve to prevent wastage: pack sizing, airtightness and re-sealing systems to prevent drying, hardening or spillage of products. A study has been initiated by the French National Packaging Committee on these subjects (CNE, 2011). Experience shows that it is often connected decisions that foster the reduction of wastage; for example, corporate tax deductions have encouraged donations to charities; an increase in the charges for effluents containing organic matter has caused companies to question their practices. In France, as part of the Grenelle II law, the announcement for 2012 of compulsorily separate collection of biowater for large producers is already a driving source for innovation.

At a retail level, use-by dates and best-before dates are very poorly understood and often mistaken: although the former, which concerns chilled products, is necessary from a microbiological food safety point of view, the latter, which applies to groceries, preserves or frozen products, only relates to the preservation of organoleptic qualities and vitamin contents. However, questions regarding changes to how these two dates are used, and which are now being raised in the UK,[8] can only be answered by modifying the regulations that standardise labelling at the European level.

In terms of final consumption within households, where wastage is considerable, campaigns to raise awareness and provide information and training are pertinent. Major retailers, community workers and consumer associations would have roles to play in this respect. Since November 2010, the French national Federation of Associations for Environmental Protection (France Nature Environnement, FNE), at the instigation of the ADEME, has devoted part of its website to the prevention of food waste, describing the savings that are possible, providing tips and advice and suggesting recipes to use up left-over food.

In a context of increasing consumption constraints (rent, subscriptions, health, transport, etc.), reducing wastage is a significant and easily accessible source of purchasing power, even if, as specified by Jakob Granit from the Stockholm International Water Institute (SIWI) when talking about the Lundqvist report (Lundqvist et al., 2008), it is also perhaps a question of culture, because in many countries it is unthinkable to eat left-over food.

With respect to catering, one option may be to overhaul the gram weights specified in the French Government Procurement Code for Catering and Nutrition (GEMRCN): 200 g of food per person and per meal are estimated to be thrown away in upper secondary school canteens, a large proportion of this being bread. The obligation to adhere to gram weights and not to increase quantities is already a first step that was provided for under the Agriculture and Fisheries Modernisation Law[9] in 2010 and the National Food Plan.

---

[8] http://www.metro.co.uk/news/861084-best-before-dates-on-food-could-be-scrapped-to-stop-waste.

[9] http://agriculture.gouv.fr/lmap.

Such measures to reduce wastage in the catering sector should be accompanied by training actions, focused on young people in particular. Part of the technology curriculum in lower secondary schools could be devoted to acquiring knowledge on environmental issues, sustainable development and wastage in general, coupled with basic courses on hygiene, balancing diets and cost management (as suggested by the Operational Committee no. 34 of the French Environment Round Table[10] and taken up by the INRA scientific report on food habits).[11] School canteen meals could also be used to support practical exercises, as an extension to theory lessons. The conclusions of the behavioural study of the *AlimAdos*[12] 2005–10 programme showed that children and teenagers are keen on these ideas. At high school level, contests between establishments have been initiated in the USA, in partnership with private contributors; these aim to encourage savings, selective collection of waste and recovery of all consumable food; prizes are awarded for reducing waste levels. The initial objective of these experiments was to combat obesity, but they have also produced results with respect to wastage. The same type of action is now expected to be taken in Europe.

Public policy should contribute to restoring less wasteful behaviour; however, before seeking to change behaviour, it is necessary to raise awareness for the 'reality' of wastage and to quantify it precisely. Most wastage is linked to the way we produce, we market and we eat food and is not perceived by those concerned who do not even realise they are throwing away goods or wasting money. Advances in preservation techniques and a sustained drop in food prices have reduced both the market and symbolic values of food.

7.5.2   Southern countries

Forty years after the FAO and its research and development partners committed to post-harvest loss reduction, it remains an ongoing problem in Southern countries. The return to today's public arena invites us to review underlying concepts of this complex issue.

Several ideas can be highlighted. First, the widespread idea that prevailed in the 1970s, according to which traditional farming and post-harvest practices were responsible for the high levels of losses, was rejected by the FAO as early as 1994 as a result of its global assessment of the many projects carried out: 'it

---

[10] http://www.legrenelle-environnement.fr/IMG/pdf/rapport_final_comop34.pdf.

[11] http://www.inra.fr/l_institut/expertise/expertises_realisees/expertise_comportements_ alimentaires.

[12] http://www.lemangeur-ocha.com/fileadmin/images/dossiers/AlimAdos-Questions-reponses. pdf.

seems that traditional technologies carried out by farmers over the years have proven useful. They are still predominantly in use today because they are well anchored in the culture of the various ethnic groups. Nevertheless, the socio-economic evolution and the ecological changes have considerably altered the production and marketing conditions for staple foods (grain and tubers). The post-harvest system must therefore be adapted to the new context, which is continuously evolving.[13] Poor compatibility between technical innovations applied to traditional practices and to local conditions has been a major source of losses at several post-harvest stages: choice of hybrid varieties vulnerable to pests; introduction of additional production seasons under less favourable meteorological conditions; frequent and uncontrolled opening of hermetically sealed grain storage containers under modified atmosphere, to cite but a few examples.

Second, consideration of the local context, including local practices and varieties, and involvement of the local population through participatory approaches, remain determining factors for successful harvest preservation.

In this context, an overview of post-harvest systems, integrated in their social, cultural, economic and political frameworks, is needed (Kitinoja *et al.*, 2011; World Bank, 2011). Among the social and cultural factors involved, gender, for example, can play an important role in labour division, for access to resources and their control, and intervenes in the choice of post-harvest operations.[14] Wide-ranging and well-documented social and human sciences studies (anthropology, ethnology, ethno-archaeology, ethno-botany, etc.) (Anderson *et al.*, 2002; Ertug, 2006; Kadim, 2009) on local knowledge of traditional societies could make a valuable contribution, but still struggle to become associated with multidisciplinary research in agronomy, process engineering or marketing and management.

Of the $940 billion that the FAO reckons is necessary for investment between now and 2050 in sub-Saharan Africa alone to fight against hunger in the region, up to half (47%) will be required in the post-harvest sector (FAO and World Bank, 2010): (1) cold and dry storage, (2) rural road infrastructures (particularly

---

[13] Source : http://www.fao.org/docrep/w1544e/W1544E07.htm#Chapter 8: Grain Quality.

[14] These factors may provide information on the functioning of post-harvest systems, the labour division between men and women, and on obstacles and opportunities relative to change. For example, the gradual introduction of communal mills, managed by men, has completely overturned the social life of women. Even though the work was hard, they are almost sorry to have abandoned manual emmer threshing, which previously favoured the development of 'mutual support networks' and fashioned the social life of women (D'Andrea, 2003). A GTZ guide focused specifically on gender-specific aspects in the post-harvest sector (Günther and Zimprich, 1997).

Box 7.2   PostCosecha: stimulating the economy by reducing losses

PostCosecha (http://www.postcosecha.net) is a rural development strategy set up by the Swiss Department for Development and Cooperation in 1980, first of all in Honduras and subsequently throughout Central America. The objective of PostCosecha was to reduce post-harvest losses of basic foods and to generate additional income for local people. It is based in particular on simple and accessible technology that is easily accepted by the population (mainly family-sized metal storage silos), and their manufacture by local tinsmiths who also provide assistance with repair and maintenance. This project benefits both households (availability of better quality food, more hygienic conditions in houses, less work for women, better market situation, higher incomes for tinsmiths) and communities (creation of employment, stabilisation of market prices, poverty reduction, improvement in living conditions) (Herrmann, 1991).

Following the success of PostCosecha (500 000 silos used in 2005), metal silos have been used successfully in 16 countries on three continents during the past 10 years (FAO, 2008b).

A similar experiment concerns triple-layer bagging of cowpea, developed by researchers at Purdue University in the USA (Baributsa *et al.*, 2010). This bag system hermetically seals the product and enables its conservation after harvest, instead of farmers being forced to sell when prices are low. The project has developed a local supply chain for the bags.

---

important to enable access to markets and collective storage systems), (3) rural and wholesale market facilities, and (4) first stage processing of raw products.

Courses of action to reduce post-harvest losses fall into two major areas: technical innovation and organisational innovation. These actions could usefully be accompanied by assessment of the current and future extent of these losses.

7.5.2.1   A need for technical innovations

Technical solutions are needed at all stages in post-harvest systems; the following are of particular importance:

- equipment to stabilise raw products (drying, salting, sweetening, smoking, fermentation, thermal treatments) that target technical efficacy in terms of yield, energy (ideally of renewable origin) and the environment, with a particular focus on the nutritional qualities and on food safety of the products obtained. These treatments could even create added value and open opportunities for new markets, for example, for export. Such equipment often does not require considerable capital expenditure and is accessible to small- and medium-sized companies and to women's groups, which are high priority targets;

- techniques and equipment for storage at family and community levels (see Box 7.2; FAO, 2008b), airtight storage and transport, improvements to packaging systems, particularly for perishable products (Manalili *et al.*, 2011), and a cold chain based on traditional systems. Storage capacity at a family level is important as it will prevent farmers from having to sell their harvest at a low price so as not to lose it to pests, and from having to purchase later for their own consumption supplies at a higher price;

- storage methods based on the use of bio-insecticides that are largely harmless to human health and financially accessible, and based on traditional or integrated pest control methods, accompanied by training in the use of these products;

- transport infrastructures, vehicles and packaging materials;

- communication infrastructures (notably mobile phones) to enable access to market information and trade.

7.5.2.2   Support from organisational innovations

Knowledge dissemination and access to capital, investment, information and markets are drivers that can reduce losses and may depend on the organisation of post-harvest operators and support services. In this respect, producer organisations or cooperatives may be particularly useful: common specifications can favour the adoption of improved practices and increase added value of products in a collaborative approach (Murthy *et al.*, 2009). Costly investments, such as refrigeration systems, can be shared (Spore, 2011).

Access to markets, whether they are domestic or export, is essential for operators to capitalise on their efforts. Effective public policies and private sector investment will be needed to support the structuring of supply chains and organisation of trade. For example, the World Food Programme's (WFP) Purchase For Progress (P4P) project (2010)[15] procures market access for the most vulnerable smallholder growers. By offering them sales opportunities for their

---

[15] http://www.wfp.org/content/mid-term-evaluation-wfp-2008–2013-%E2%80%9Cpurchase-progress%E2%80%9D-pilot-project-terms-reference.

cereal harvests (guaranteed contracts for a three-year period), investment in post-harvest infrastructure can be made in a future perspective.

Another essential element is access to credits for material investments to smallholder farmers and to small market operators. A system providing access to capital is described in Box 7.3.

Box 7.3   Financing trade with agricultural commodities by Warehouse Receipt Financing or Inventory Credit

When African, Asian and Latin American countries, and more recently Eastern European countries, agreed to liberalise their economies, agricultural markets were also affected. Previously, state agencies were responsible for purchasing and warehousing harvests between the production season and the gap period. In the course of liberalisation, state organisations did no longer assume this purchasing and warehousing role. New types of organisation of agricultural trade therefore became essential.

However, private stakeholders have difficulties in accessing the necessary credits for financing agricultural trade (as this is not really considered as a guarantee in most of the countries concerned). Warehouse receipt financing, a longstanding method for access to funds that is practised in many countries, has been suggested by the FAO, the World Bank and the Natural Resources Institute (NRI) as one solution to encourage improved practices and improved storage facilities in order to foster agricultural development and dynamic trade (Coulter, 2010; Giovannucci et al., 2000). Three parties are involved: the financier, the depositor and the warehouse operator. The principle is that the depositor places a certain volume of his crop in the warehouse and receives a receipt from the operator based on the volume and the quality of his product certifying this deposit. This receipt equals a guarantee that enables the depositor to obtain credit from the financier.

Warehouse receipt financing has been shown to be effective in reducing storage losses and provides a complementary system to storage on the farm. According to experience in Madagascar, Mali and Niger, improved production technologies have thus been adopted more easily by smallholder farmers (Coulter, 2010). Based on experience in several countries, Coulter and Shepherd (2001) discussed the advantages, drawbacks and elements for the success of this system in general.

7.6   Experimentation in Southern countries: models for the North?

At all times and in all cultures, organic waste, crop residues and products unfit for human consumption have been recycled as livestock feed among poor populations, and particularly (but not only) in difficult environmental contexts, in both Northern and Southern countries.[16] In return, livestock supplies fertiliser, labour and a source of nutritious foods high in proteins and micronutrients, an exchange currency and acts as 'living capital' as insurance against any future disasters or unexpected events (Faye, 2011; Faye and Duteurtre, 2009). Complementary functions between humans, plants and animals, which are fundamental to mixed crop–livestock systems, are a source of productivity for food systems and have been the subject of considerable research (Herrero et al., 2010). For example, studies have focused on improving dual purpose cereal varieties (maize, wheat, sorghum, millet): to improve the use of crop residues as livestock feed without compromising grain yields serving human consumption (Herrero et al., 2010).

Integrated systems that include aquaculture are particularly effective in using organic matter. The VAC system,[17] which has its origins in Vietnam, is the integrated system most frequently cited in the literature, but other systems based on the same principle also exist. The underlying principle is that there is no final waste output: the waste from one system becomes the input of another, and nutrients are entirely recycled.

Anaerobic digestion, or methanisation, is another way to treat organic waste and livestock manure. In the same way as with integrated systems, countries in South and South-East Asia, and notably India, have been using biogas systems for a long time, and their development stepped up from the 1970s (Barnett et al., 1978), particularly regarding the family use of biogas (e.g. supported by the Khadi Village Industries Commission (KVIC) as early as 1954). Apart from energy

---

[16] According to the FAO, two-thirds of rural households throughout the world raise livestock as a means of subsistence for more than a billion people (FAO, 2009a).

[17] The acronym VAC refers to three words: V for *Vuon* (vegetable garden, and in the broader sense any type of plant cultivated in the garden, field or forest), A for *Ao* (pond, and any type of aquaculture in fresh, brackish or salt water) and C for *Chuong* (enclosure for animals, meaning the raising of livestock and poultry, as well as birds and bees). The most frequently cited VAC comprises vegetables–fish–pigs. Numerous variations of integrated systems, sometimes combining the raising of other domestic animals, have been described (Kumar and Ayyappan, 1998): fish–duck, fish–rabbit, fish–goat, fish–poultry and, in particular, fish–silkworm–mulberry. The fish–rice system has been known in India and South-East Asia for 1500 years, but is now in decline because of the use of pesticides in rice production that proved harmful to fish.

recovery from biomass, the digestate, a moist product rich in organic matter, can be used as a fertiliser (Balasubramanian and Bai, 1992; Gopal *et al.*, 1996).

Renewed interest in mixed or integrated farming systems and in recovery options of organic waste can be observed in both Northern and Southern countries. Yet when recovery options themselves become a final aim for agricultural biomass production, this raises the question of competition between different uses of biomass and agriculture's primary function of food production.

These different examples illustrate the importance of placing the issue of food losses and wastage in the widened context of systemic analysis on production and utilisation of agricultural biomass in general (see Chapter 4). The concept of closed-loop supply chain models inspired by natural ecosystems where 'losses', waste and effluents generated by one system become an input for another system (Fan *et al.*, 2006) could be an important conceptual contribution to rendering food systems more sustainable.

7.7   Questions for research

Over and above the general need for data on losses and wastage, several research options emerge from the above review.

7.7.1   *Northern and Southern countries*

7.7.1.1   A need for knowledge and quantification methods

In both Northern and Southern countries, food losses and wastage are very poorly understood, except for some studies conducted in the leading countries on the subject (UK, Belgium, Sweden, USA) and with the exception of the database on post-harvest losses in sub-Saharan Africa. Most countries require not only quantitative data, but also methods for data collection, particularly in countries that do not have any reliable statistics system. The aim should be to develop methods that provide rapid and inexpensive access to representative data on losses and wastage.

7.7.1.2   Definitions

The definitions and concepts underlying food losses and wastage make no distinction between losses of food in the strict sense of the term and different recovery options (human food, animal feed, energy, chemistry, fertilisers) as co-products, by-products and waste, although food that no longer fulfils its initial purpose, i.e. feeding people, can follow other recovery pathways. How could food losses and wastage be defined so as to reflect the complexity of production and recovery systems for organic matter? These issues are of technical, statistical and regulatory importance.

7.7.2   Northern countries: reducing wastage

Wastage is always the result of compromises within different constraints. Wastage has to do with safety, economic evaluation and social problems and is a source of environmental impacts (air, soil and water pollution). How are these choices being made within different supply chains as well as in households? Which dimensions are taken into account in decision-making, and which are not? Which decision-making tools can support the complex reasoning of these dimensions?

Without jeopardising food safety and consumer health, rapid, accurate and accessible tests which still need to be developed could help the decision process at a retail level by determining the microbiological quality of foods.

In terms of relationships between technologies, packaging and food losses (e.g. LCA of packaged products simultaneously analysing impacts of packaging and impacts of food production: land, water, inputs, labour, etc.), which modelling method should be used to determine combined impacts?

Which decision-making tools should be made available to consumers and operators?

Which industrial processing technologies could be applied, so as to make better use of raw materials and to increase the proportion that is edible by humans? Integrative research for eco-designing processes that integrate loss reduction, better use of by-products, optimisation of packaging and transport systems and logistical parameters for transport, etc. is needed. Better use of raw materials also requires innovation in the design and conception of food products that will be acceptable to consumers.

What are the drivers behind changes in attitudes and values linked to food resulting into greater wastage?

Which stakeholder interests go against a reduction of losses and wastage? Which negative effects may be associated with reducing wastage, and to what extent (number of jobs associated with wastage, economic value of waste management)?

What contribution can knowledge on recycling and recovery strategies in Southern countries make in the context of Northern countries' food systems? VAC system or VAC principle? Which spatial scales are relevant? What is the economic viability of systems, and according to which societal parameters (rise in the price of fossil fuels, tightening of environmental regulations, etc.)? How can stakeholders from different sectors be mobilised?

7.7.3   Southern countries: reducing post-harvest losses

- Research is necessary to characterise the huge number of local varieties of food-producing crops and their associated post-harvest operations, by product and by country. The idea is to describe the biological and technological specificities of local varieties in order to optimise post-harvest systems and thus reduce losses. This knowledge still needs to be obtained, but some information is already available from the human and social sciences and should be mobilised.

- Which varieties are best adapted to which use? Improved dual-purpose crop varieties (grain for humans, residues for livestock) can optimise mixed farming systems. Research on strategies to make best use of co-products and by-products of any agricultural commodity could already lead to the design of zero-loss systems. There is also an obvious need for optimisation and control of product stabilisation processes.

- Which technical options (cold chain, storage, packaging, transport) are the most effective in reducing post-harvest losses? Which obstacles and which opportunities exist relative to the local context, for example, the existence of marketing opportunities, organisation of supply chains, support from social and cultural traditions, access to energy? What environmental, economic and social impacts could be expected from these technical options? Is there any conflict between these impacts? How can these choices be managed, using which tools? Analyses of returns on investment are needed: is there an acceptable loss level?

- What traditional and modern innovative strategies used to prevent losses can be observed amongst populations' practices? What product recovery and recycling strategies could be implemented? Research should aim to describe different practices to make use of inedible products or to preserve their edibility.

- What are the impacts linked to the transformation of food markets alongside the industrialisation of supply chains and penetration of retailing companies? What are the effects of tightened public and private quality standards introduced on both domestic and export markets? Will these give rise to even more losses? Will they foster the expansion of parallel distribution channels? What health effects might appear? How can smallholder farmers and small-scale post-harvest operators comply with these standards? On the other hand, under which conditions might the introduction of standards help to reduce losses?

- In cities, in particular, which changes to domestic practices and values can be observed in middle-class and affluent households, and what are their impacts on losses? Is there any increase in wastage?

- Which types of organisation between operators help to reduce losses (for example, producer organisations or cooperatives, warehouse receipt financing (see Box 7.3))? What is the role of quality standards or of market access for successful loss reduction within these systems?

7.8   Conclusions

The subject of losses and wastage is new in the field of research. Until now, it has been a fundamental notion of household management and of on-farm harvest protection. However, faced with the colossal quantities concerned and considering the impacts of losses and wastage on the entire biosphere, it has once again become a priority in efforts to achieve global food security and goes well beyond the scope of home economics and storage protection.

We have seen that even the different terms used are a controversial issue, which does neither facilitate comparisons nor the measurement of any progress in reducing losses and wastage. This of course raises questions concerning standards and regulations, but also regarding data collection, their validation and methods to limit losses and wastage, not only at the household level but in macroeconomic terms.

Recovery of this 'organic waste' is an issue associated with food losses and wastage that must be addressed in any analysis of the production and use of agricultural biomass as a whole.

Through these aspects, this chapter therefore offers a novel approach: a critical review of the definitions and concepts relative to losses and wastage, possible recovery options and a systemic analysis of the flows of agricultural biomass, and multidisciplinary research questions that stretch far beyond the search for loss reduction solutions.

However, before we can speak of food sustainability, a 'revolution' of attitudes is necessary: considering food losses and wastage as abnormal is a notion that needs to be rediscovered both individually and collectively.

8

International trade, price volatility and standards for sustainability

Authors: SÉBASTIEN JEAN, NICOLAS BRICAS
AND CHRISTOPHE GOUEL

Contributors: JEAN-CHRISTOPHE BUREAU,
ARMELLE CHAMPENOIS, BENOÎT DAVIRON,
ALEXANDRE GOHIN AND ÉLODIE MAITRE D'HÔTEL

Questions of food sustainability cannot be limited to the national sphere, insofar as international interactions are undeniable. This chapter focuses on their role, concentrating on the volatility of global food prices and on public and private norms and standards.

The international dimension of these questions is reflected first of all by questions concerning the volatility of agricultural prices. Although volatility itself is not the preserve of international markets, it has become difficult to address this issue without taking account of the global dimension of agricultural markets, in terms of both the formation of prices and their subsequent adjustments. These issues have become particularly acute since the recent crises affecting food prices. When linked to the issue of food sustainability, they lead to questions on the types of adjustments made at a global level and the resulting differences according to the level of development of a country, as well as justifications for, and the modes of, public intervention. Commercial policies that are directly affected by international interactions play a particular role in this respect, and this merits discussion in order to analyse the challenges inherent in their application and any disciplines associated with their use.

However, international interactions are not restricted to prices. Indeed, trade affects the capacity of public and private agents to lay down and subsequently

Food System Sustainability: Insights from duALIne, eds. Catherine Esnouf, Marie Russel and Nicolas Bricas. Published by Cambridge University Press. © Cambridge University Press 2013.

apply norms and standards, as well as the scope of their influence; in return, these norms and standards may exert a substantial effect on trade. During the past 20 years, the rapid development of public and private norms and standards linked to sustainability issues has thus raised a series of questions on the reasons for them and their consequences in terms of international trade.

8.1   Adjusting to the volatility of global prices

The openness of food product markets may appear to be limited when compared with the number of products manufactured: in 2010, exports accounted for about 7% of the world production of brown rice, 11% of maize and 19% of wheat.[1] Nonetheless, because such products are generally homogeneous, these percentages are sufficient to consider that the supply–demand balance is established at a global level. For this reason, even if differences linked to quality or the site of delivery are seen, there is an unquestionable international market for these goods, and price changes above all reflect imbalances between supply and demand in this market. It is thus possible to refer to a world price, although this raises the question of determining who ultimately makes the necessary adjustment. If agricultural prices are so volatile, it is precisely because consumption reacts little to price. The principal factor for adjustment in developed countries lies in changes to carry-over stocks. It is only when these stocks reach a minimum level that other consumptions are adjusted and prices become both high and volatile. It is customary to consider that uses not directly devoted to human food consumption are more elastic than food uses. For example, this is the case for animal feed, which can more easily switch between cereals than can consumers. However, support policies for biofuels may lead to a major qualitative change in this respect. Indeed, because of compulsory incorporation requirements ('mandates') concerning a minimum content of agrofuel in fuels of fossil origin, the demand for agrofuels is particularly inelastic to price and causes an adjustment in other uses.

In developing countries, the situation may be slightly different. First of all, the correlation between production and consumption is stronger than in developed countries, emphasising the lower smoothing effect of stocks and trade. Storage is indeed more expensive in these countries, because of unsuitable technologies and/or unfavourable climatic conditions. Carry-over stocks are smaller in many African countries, and the lean period during which stocks from the previous harvest are becoming exhausted and the new harvest is

---

[1] Figures calculated from the Production, Supply, and Distribution Database, Foreign Agricultural Service, US Department of Agriculture (http://www.fas.usda.gov/psdonline/).

not yet available may occasion very marked price increases. Because of low carry-over stocks and less non-food use in the poorest countries, most adjustments are driven by either trade or human consumption. The effect of price volatility on food supplies is therefore much more direct than in richer countries. The FAO thus estimates that in 2007, price rises increased the number of undernourished people in developing countries by 75 million when compared with the average for the period 2003–05 (FAO 2009b, based on evaluations by the World Food Programme, which qualitatively confirm these results).[2] Even within poor countries, however, this effect is markedly heterogeneous. It may notably depend on sources of income and also – among other factors – on levels of savings, on institutions, on government action, on the place of residence and on the formal or informal solidarity mechanisms that may be available. There is no doubt that these determinants of food vulnerability also merit further study, insofar as targeting the populations concerned is often determinant in the efficiency of support programmes (Barrett, 2010).

8.1.1   *The role of international trade*

International trade plays a particular role. It is often perceived as a destabilising force as events external to local production and consumption conditions may, through its intermediary, have dramatic effects on domestic prices. Politically, in many cases, it may also constitute a useful scapegoat for national governments.

Nevertheless, trade can spread yield shocks across a wider market and thus share the risks of agricultural production. For this reason, production hazards are much lower at a global scale than at a national scale (Wright, 2009), as can be seen from variance or extreme episodes. To illustrate this effect, Figure 8.1 shows the coefficient of variations in rice and wheat yields by country and throughout the world for the period 1960–2010; this calculation was made following extraction of the evolution trend for yields. By decreasing order from left to right, this graph covers the 10 largest producers in the world in 2010, excluding the former USSR. In both cases, no major producer presented variations in yield nearly as low as the world taken as a whole. For example, with respect to wheat in the EU-15, the coefficient of variation was approximately 50% higher than that of the world as a whole.

---

[2] Although it is the most widely used, the FAO estimate of the undernourished population, based on food data at a national level and on strong hypotheses on its distribution within each country, should be considered with caution and may differ significantly from other evaluation methods (Barrett, 2010). Nevertheless, there is no doubt as to the qualitative results.

(a)

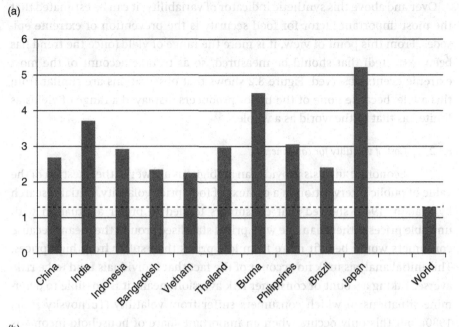

(b)

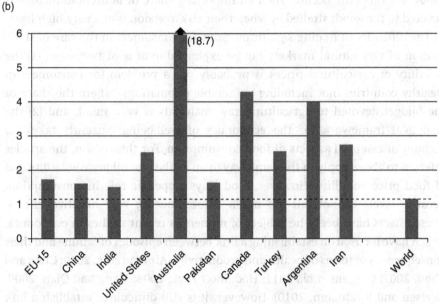

Figure 8.1 Coefficients of yield variations, by country and worldwide (1960–2010) for (a) rice and (b) wheat (from the Production, Supply, and Distribution Database, Foreign Agricultural Service, US Department of Agriculture). Trend extracted using the Hodrick–Prescott filter, parameterised at 400.

Over and above this synthetic indicator of variability, it can be estimated that the most important factor for food security is the prevention of extreme episodes. From this point of view, it is more the range of yield (once the trend has been extracted) that should be measured, so as to take account of the most extreme events observed. Figure 8.2 shows that observations are similar from this angle, because none of the major producers displayed a range of yields as limited as that of the world as a whole.

8.1.2   Cost of volatility for consumers

Economic analysis provides an ambiguous answer to the question of the value of public intervention in a context of food price volatility. Initial research by Waugh (1944) showed that consumers tended to prefer a situation with unstable prices rather than one with prices stabilised around the mean, because consumers would benefit more from low prices than suffer from high prices. This initial analysis took no account of the fact that individuals tend to be risk-averse. Taking account of consumer risk aversion means it is possible to determine situations in which consumers suffer from volatility (Turnovsky et al., 1980), but this only occurs when an important share of household income is devoted to the goods studied or when their risk aversion is at a very high level.

The difficulty of finding significant gains for consumers in the case of stabilisation of agricultural markets can be explained in at least two ways: (1) the volatility of agricultural prices is probably not a problem for consumers in wealthy countries (not including vulnerable populations) where the share of the budget devoted to agricultural raw materials is very small, and (2) the analytical framework for the economics of well-being currently takes no account of essential aspects of food consumption. For this reason, the articles referred to above apply in the same way to an analysis of oil price volatility and of food price volatility. However, food plays a specific role in consumption, because levels cannot fall too much before causing very negative effects. These effects have been the subject of numerous recent studies in economics, which have focused on establishing a link between episodes of famine and their consequences in terms of health and productivity (Almond et al., 2010; Chen and Zhou, 2007; Gorgens et al., 2011; Hoddinott et al., 2008; Meng and Qian, 2009; Neelsen and Stratmann, 2010). However, it is still difficult to establish a link between the consequences of undernutrition and their economic impact.

8.1.3   Justifications for public intervention

The fact that volatility is costly to consumers does not necessarily justify public intervention; another prerequisite is that public intervention may be welfare-enhancing, so it is necessary to try and prove the existence of market

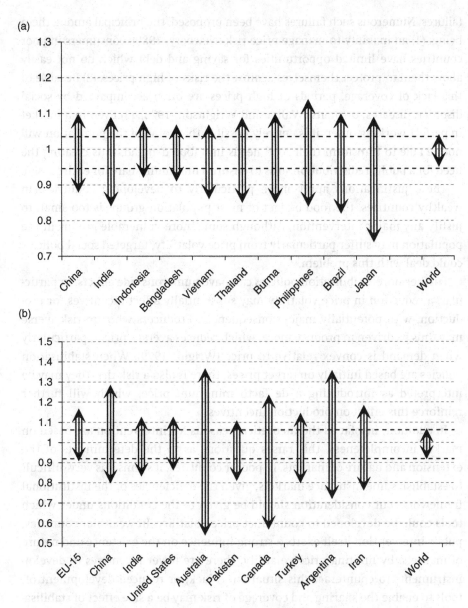

Figure 8.2 Range of variations in yields, by country and for the world as a whole (1960–2010) for (a) rice and (b) wheat (from the Production, Supply, and Distribution Database, Foreign Agricultural Service, US Department of Agriculture). Trend extracted using the Hodrick–Prescott filter, parameterised at 400.

failures. Numerous such failures have been proposed, the principal among them being: (1) an inability to manage price risk (Newbery, 1989): consumers in poor countries have limited opportunities for saving and debt which do not easily allow them to protect themselves against the risks of high prices. (2) Because of this lack of coverage, periods of high prices are often accompanied by social disorder, diseases, or thefts (Walter and Wrightson, 1976). (3) Even in a market in perfect working order, it is entirely possible that some of the population will not be able to transform their food needs into food demand, thus creating the need for a public redistribution so as to ensure a minimum share for each person.

These justifications apply more particularly to developing countries. In wealthy countries, the food budget of most population groups is too small to justify any major intervention. Although some more vulnerable groups in the population may suffer particularly from price volatility, targeted social policies could deal with this problem.

Furthermore, stabilisation policies can have numerous side effects. In particular, a reduction in price volatility may substantially affect incentives for production, with potentially major consequences. Producers who are risk-averse may thus be driven to produce more, which reduces average prices, particularly when demand is convex relative to price (Wright, 1979). When stabilisation policies are based initially on target prices, there is also a risk that they may be interpreted as introducing a de facto minimum price, which will further reinforce this effect on production incentives.

In addition, insofar as the theoretical reasons for intervention are based on market incompleteness, this raises questions as to the determinants of the extension and nature of markets. In poorer countries, it seems it is very difficult to establish some insurance markets, given the deficiencies of the institutional framework, but consideration should be given to the conditions under which tools could be developed to counteract price volatility. In wealthier countries, public intervention itself exerts a strong influence on the existence and nature of markets: by limiting price volatility, there are fewer incentives to develop instruments to counteract this situation. A lack, or reduced development of, tools to enable the sharing and coverage of risk may be a side effect of stabilisation policies.

8.2   Management of food price volatility: policies and food sustainability

The justification of public intervention regarding the management of agricultural risks remains generally controversial at an academic level. Nevertheless, such intervention does exist in many countries, which raises a certain number of questions relative to their management. In the context of the soaring agricultural and food prices in 2007–08, and the subsequent food crises, these questions took on particular importance (von Braun *et al.*, 2008). There has been much debate about the source of this volatility, the justification of public intervention, the types of intervention that should be prioritised according to their objectives (to reduce the volatility of food prices or limit its effects), their mode of governance (regulation by the market or by governments) or their level (national, regional or even local), and finally regarding the levels of volatility that might be considered as being acceptable or excessive.

In practice, in recent years there has been a certain revival in developing countries of policies to manage the volatility of food prices (Abbott, 2010), which has raised two types of questions. The first concerns the conditions for the efficiency of policies to manage food price volatility in developing countries, while the second relates to the compatibility of national policies to manage food price volatility with international trade regulations.

We lack hindsight in developing countries as concerns the greater public intervention intended to limit the volatility of food prices, as in many cases these interventions were put in place in response to the food crises of 2007–08 and 2010–11. It does however appear that the results were contrasting in nature, as in some cases reinforced intervention reduced price volatility (Madagascar 2008), but in others it had no effect on volatility (Mali 2005, Mali 2008), or even worsened the situation (Madagascar 2004, Kenya 2008). These contrasts raise questions as to the conditions required for these policies to be effective. Some case studies have shown that the institutional context in which these policies are applied is a determining factor in their efficiency. The following elements seem to be particularly crucial:

- The ability of the State to manage the costs linked to implementing intervention, which are particularly high in the case of managing buffer stocks (Gérard *et al.*, 2011), or the introduction of innovative systems to enable the sharing of these costs; for example through temporary public grants for private storage facilities (Baris and Cordier, 2011).
- The ability of the State to apply its intervention policies in a foreseeable manner so that private operators can appropriately anticipate government action and position themselves accordingly in the markets, particularly if they are active in external trade (Byerlee *et al.*, 2006; Chapoto and Jayne, 2010; Tschirley and Jayne, 2010).
- The ability of the State to ensure the implementation of and compliance with intervention policies, which may be crucial in the case of border measures, for example (Gérard *et al.*, 2011). Indeed, in many developing

countries, informal trade may account for a large share of international trade, which is likely to markedly restrict the efficiency of domestic policies, which may not be coordinated at a regional level (Tschirley and Jayne, 2010).

- The ability of the State and private operators to draw on good quality expertise, that can then drive political choices, whether these choices concern the introduction of policies linked to external trade (definition of tariff measures to control imports and exports) or to the domestic market (definition of optimum stock levels and price levels that trigger intervention operations) (Timmer, 2010).

The development of consultation structures between public and private operators is also mentioned as an element that will favour the factors listed above (Fouilleux, 2010). As well as these case studies, the relative importance of these conditions to the ability of policies to limit food price volatility still needs to be determined.

8.3   Use of trade policies to cope with price volatility

Although trade policy interventions have been widely discussed in the context of the recent crises affecting agricultural prices, they are in fact common. Based on an estimate of the nominal level of support provided for different sectors of agriculture, Anderson and Nelgen (2012) were able to show that border measures were negatively correlated to the evolution of world prices. Figure 8.3 illustrates the example of rice in South-East Asia, but this type of relationship is quite commonly observed. The link is significant, in both industrial and developing countries; it applied in the 1970s and it still applies today. The result is that restrictions or incentives (usually, but not always, tax related) are applied to exports or imports. To understand their implementation, two cases should be distinguished.

The most classic case is what can be qualified as a situation of glut. This reflects a period of relatively low prices, or more generally a context in which emphasis is placed on competition between exporters for market access. Interventions meet the demand for protection from producers and lead to the use of a combination of import restrictions and export incentives. The link with price volatility may be more or less clear, depending on the case, but it notably becomes evident when these measures are adopted to support a guaranteed minimum price mechanism. The best illustration of this is probably the European policy that applied prior to implementation of the Marrakesh Agreement, when intervention prices were sustained using variable import

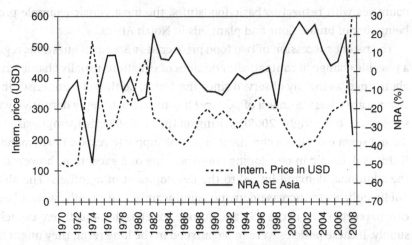

Figure 8.3 Internal price and level of support provided by trade policies: the example of rice in South-East Asia (1970–2008) (from Anderson and Nelgen, 2012). Note: the solid line represents the nominal level of support for the sector via the intermediary of trade policies. The dotted line represents the internal price in US dollars.

duties and export subsidies. Systems of price bands were also employed in some countries, particularly in Latin America. More generally, it is clear that a certain number of countries selectively adjust their degrees of customs protection, taking account of world price levels and prospects.

Concerns regarding food security lead us to highlight another type of case, which, on the contrary, corresponds to situations of scarcity. These episodes are marked by high prices, or more generally by competition between importers to secure supplies at moderate prices. In this case, it is necessary to respond to a demand for protection from consumers, and particularly the most vulnerable groups. Trade policies can be used in this context through restrictions on exports or incentives for imports. Measures that affect exports have been seen to exert a particularly strong influence, because some of them have had an abrupt and considerable impact on international markets. This was notably the case of the export restrictions implemented by Thailand with respect to rice in 1973, by India, Vietnam and China on rice in 2007–08, and by Russia and Ukraine on wheat in 2010–11. However, measures that promote imports are also widespread. The frenetic purchasing of rice by the Philippine authorities in order to increase their stocks in the autumn of 2007 is an example of the potentially sudden nature of these measures (Timmer, 2010). Furthermore, Demeke *et al.* (2009) estimated that 25 countries reduced their customs duties on basic foodstuffs in 2007–08. Direct or indirect support for imports is inherent in the guaranteed price systems applied by a certain number of developing countries with respect to basic foodstuffs, the most classic example probably being that of bread, flour and plant oils in North Africa.

The recent succession of two food price crises also raises questions regarding a possible change of context: after decades of situations usually characterised by abundance, as mainly observed since the 1980s, do these recent crises indicate that we are entering an era of penury? It is too early to reply definitively to this suggestion, but Wright (2009), writing on the 2007–08 crisis, emphasised that it did not seem to reflect a chronic inability of supply to respond to demand, even if this is dynamic in developing countries: the one exception, however, being the additional demand linked to the development of agrofuels. The ability of public policies to increase demand, significantly and rapidly, as has been the case in recent years in the USA and Europe, exceeds the adjustment capacities of supply. If these policies were to be pursued in the long term, they might indeed create shortages, either via a direct effect of demand on certain food products, or via an indirect effect linked to competition for land use. A correlative question concerns entry into a phase of increased world price volatility. But once again, analysis of the data currently available in this respect does not enable a conclusion as to such a long-term rise, despite recent episodes. Indeed, Gilbert and Morgan (2010) underlined the fact that, as a general rule, the volatility of agricultural prices was less marked during the past 20 years than previously. If volatility has tended to increase recently, this is not incoherent with historical observations, except regarding rice. Their main conclusion was that it is too early to draw any clear lessons, which means that new research in this area will be valuable when more experience has been gained.

Whatever the context, trade policies have some unquestionable benefits when it comes to controlling price volatility. They are relatively simple to implement, and substantial and rapid results can be expected. For example, the stabilisation of internal prices within the European Union until 1992 was unquestionable, and countries that markedly restricted their exports generally succeeded well in isolating themselves from the world market. A stock policy requires anticipated action and the payment of storage costs, and its extent is limited by stock levels; the same does not apply to trade policies. However, the drawbacks of trade policies are also considerable; their tendency to distort the market leads to inefficiency costs; by being isolating from market signals, their systematic use creates problems of resource allocation, promoting production in a context of low world prices, and tending to reduce it in the opposite case. Finally, their implementation may be problematic, either because this gives rise to fraud or avoidance, or because the policies are unsuitable in terms of their design and/or the information available on them (Jayne and Tschirley, 2009). The analysis of policies and the establishment of optimal regulations are important areas for research in this respect. In addition, although they can indeed help to prevent episodes of extreme prices, trade policies also have very marked redistributive effects relative to the protection they provide (Gouel and Jean, 2011). For example, restrictions on exports induce a considerable transfer of wealth (in relative terms) from producers to consumers. From this standpoint, targeted measures on the agents requiring protection are preferable.

In terms of international governance, however, the main problem posed by the use of trade policies to tackle price volatility is their non-cooperative nature. In all cases, this type of intervention leads to partial or complete isolation from global market fluctuations. This then limits the role that the country will play in adjusting the underlying imbalance between supply and demand. In addition, indirect effects on global prices counteract direct effects on internal prices, which diminishes the efficiency of such measures in a large country, and might even cancel them out in extreme cases (Martin and Anderson, 2011). More generally, the outcome is negative regarding relations between different countries, as the measures adopted by one may operate to the detriment of others.

This risk has long been identified regarding competition between exporters in a context of low world prices. The multilateral trade system, designed at a period marked by the trauma of the trade wars of the 1930s and then dominated by mercantilist concerns, was indeed intended to limit this competition. This explains why strict regulations surround the use of import restrictions and export grants. In a context of high prices and competition for supply, on the other hand, the multilateral trade system seems to be relatively helpless. Thus, no strict regulations have been determined regarding grants for imports. As for restrictions on exports, they, like other measures, are subject to a ban on the use of quantitative restrictions (GATT, Article XI.1), but exceptions are notably allowed 'to prevent a critical situation due to a shortage of food products or other essentials' (Article XI.2.a) and to ensure price stabilisation (Article XX.i, targeting processing industries).

In this context, there have been demands for stricter regulations on export restrictions. In fact, the subject merits further investigation, although its framework should not be limited. This issue does not only concern restrictions on exports, but also subsidies (or other incentives) for imports and, in addition, trade policies in a context of shortage. The debate should also take account of the specificities of the poorest countries, as it is very clear that the question differs markedly depending on the share of food costs in the global budget, the ability of agents to protect themselves, and any credible alternatives that the government might implement. Studies should also focus on the effects of these measures on the development of such periods of shortages.

This is a major challenge for the multilateral trade system, because the political stakes are often very high in a context of shortage and of growing uncertainty linked to the challenges of climate change, to such an extent that it is difficult to make any commitment made by the governments of Member States appear credible. Furthermore, the current system for the resolution of disputes is based on a relatively short timescale when compared with the general pace of disputes under international law, which partly explains its success; however, because food crises can appear very abruptly, the procedures are much too lengthy to be able to claim to be effective with sufficient rapidity. Although cycles of low prices are often long, high prices usually result in abrupt and brutal crises, in terms of both their genesis and their effects. By its very nature, trade induces reciprocal dependence; the credibility of the commitments of trade partners is thus an important subject that merits further research.

8.4   Non-tariff barriers, standards for sustainability

The interactions inherent in trade relations go far beyond issues of price; they also influence the ability of public and private agents to lay down standards and ensure compliance with them. For this reason, recent decades have seen a proliferation of quality standards for agricultural raw materials and food products traded on international markets (it should be pointed out that we are using the term 'standard' in a very broad sense, because in administrative language, a standard is voluntary, unlike a technical regulation that is compulsory). These norms or standards initially concerned the safety quality of products, notably with respect to pesticide residues and mycotoxin contents. These standards were first of all public, before becoming national and then international (the *Codex Alimentarius*, containing scientific results that have been references since 1995 in the resolution of trade disputes, is a good example). Although some standards are compulsory, a growing corpus of optional procedures, often arising from private consortia, has emerged. And because they are imposed by industry and distributors, these standards have de facto become those of producers in many countries.

Public standards may provide new means for countries or regions to regulate their imports and protect their production. The trade barriers that result from public standards have been illustrated in numerous reports (OECD, 1999). In many cases, however, the border is blurred between an openly protectionist use of these public regulations, and the indirect negative effects on trade that result from protecting consumers or flora and fauna. It was to establish transparent rules on this point that agreements on health and plant health measures

(Sanitary and Phytosanitary or SPS measures) and on technical barriers to trade (TBT) were included in the Marrakesh Agreement. These agreements imposed not only scientific evidence (this point has since been clearly affirmed by jurisprudence) but also data supported by scientific results to justify standards that overstepped acceptable international references (such as, for example, the *Codex Alimentarius*) as a framework for legitimate practices. The aim of these SPS and TBT agreements was to prevent arbitrary measures that would be used purely in order to restrict trade.

Private actors have introduced their own standards, often initiated for reasons of industrial conformity, to organise a supply chain and guarantee health safety or even to limit legal risks in the event of a health problem that incriminates their responsibility (Barrientos *et al.*, 2001; Fulponi, 2006; Meuwissen *et al.*, 2003; Nadvi and Waltring, 2003). These private standards also allow companies to implement coherent quality procedures and may contribute to promoting international trade, particularly regarding the trade of intermediate goods. Nevertheless, these private standards can also cause the exclusion of certain producers, and sometimes certain countries, from the market. Such private stipulations are not covered by an institutional framework such as the WTO to enable the resolution of any disputes.

Alongside these mainly technical private standards, the 1990s saw a proliferation of initiatives to establish standards defined with reference to sustainable development, of immediate international scope. Private stakeholders drove this trend by means of company standards (good practice codes, directives for suppliers, principles for sustainable business management practices, etc.). Major, often international, corporations have also had recourse to recognised and reputed NGOs to testify as to their awareness of social and environmental issues. In parallel, the definition of sustainability standards in the context of multi-stakeholder initiatives started in the mid-1990s with the Forest Stewardship Council (established 1993) and the Marine Stewardship Council (established 1997). This trend accelerated during the 2000s, with the appearance of several initiatives – often qualified as 'round tables' – focused on specific crops (soybean, palm oil, cotton, sugar cane, coffee, tobacco, etc.) and based on dialogue, partnership, certification and labelling (Segerlund, 2010). These initiatives organised discussions (with a view to defining standards) among a variety of stakeholders with supposedly divergent views (Fransen and Kolk, 2007; Pattberg, 2005). The companies driving these initiatives then imposed these principles on their numerous suppliers. Thus the Euro Retail Group (EUREP), a network of European retailers has developed a private standard of this type on good farming practices (GlobalGap) and now imposes it on a large number of suppliers throughout the world; more than 102 000 producers in

108 countries were thus certified with reference to this standard in 2010. GlobalGap has thus become a benchmark regulation for the international trade of food products (Neilson and Pritchard, 2009). Requirements for food traceability are also developing. The European Union is particularly advanced in this respect, both in terms of public requirements driven by consumer movements and relative to the demands of distributors, who often go beyond public standards in order to rapidly circumscribe any potential problems and also protect themselves against any legal risks.

Overall, both public and private standards participate in a segmentation of countries and hence selective protectionism, even when the initial aim was to protect consumers (which is not a matter for debate and enables fair trade), or protect plants and animals. This has affected the poorest countries, where infrastructures, skills and administrative management find it difficult to comply with the standards required within the EU. Thus safety requirements (Hazard Analysis Critical Control Point or HACCP certification) or the capacity to react to a sanitary crisis (a condition for the export of animal products in developed countries) have in fact placed a limit on the number of countries that are able to export goods to Europe or the USA. With the reinforcement of traceability requirements, this effect has been amplified by private standards. There is thus a marked contrast between a group of Latin American countries that have been able to respond rapidly to public European demands, and a broad range of other countries, ranging from Moldova to Mauritania, which, despite no customs duties and specific technical assistance, cannot hope to export animal products to the EU. In Chile, for example, stakeholders quite frequently acknowledge that the imposition of very strict quality and traceability standards by the European Union and Japan has had a structuring effect on trade and has facilitated their exports to other countries.

8.5   Questions for research

In light of this review, numerous questions still need to be addressed in order to better understand the interactions between international trade and food sustainability.

With respect to price volatility, although the causes of the recent rise in food prices have been widely studied, its consequences nevertheless merit further analysis. In particular, it seems necessary to broaden the analytical framework on the costs of instability to integrate its effects on health (linked to both the quantitative and qualitative aspects of food) and on education, or even social and political life. Such an analysis would concern not only the costs of instability but also market failures in terms of these different criteria. This analysis could also be extended to address development trajectories and the modalities of price transmission.

Public interventions, when they are decided upon, also merit further study to better understand the conditions for their efficiency, in terms of both institutional structures and the tools employed. The consequences of these interventions also raise questions as to their influence on the development, or even the existence, of risk-sharing markets. If the institutional framework and types of markets were considered as exogenous, this could weigh on many of the consequences of introducing or suppressing public intervention.

In particular, the use of commercial policies for stabilisation purposes raises questions of international governance. Despite the potentially high costs of a lack of coordination, no sensible multilateral discipline exists in a context of shortage. A first question thus concerns the duration of this context of scarcity; recent episodes have not yet enabled any conclusions regarding a paradigm change, but this hypothesis should not be excluded, and at least the reference framework for analyses should be rethought by exploring new instruments. The political consequences of such situations also merit further study; in particular, although restrictions on exports have been the subject of sustained attention, this is not the case for incentive measures in favour of imports, which nevertheless play a symmetrical role that is far from negligible. As for the multilateral disciplines that could be envisaged, until now the question has mainly been put in political terms, but has been little studied from an economic point of view. A first theme of importance in this respect is that of the asymmetry between rich and poor countries, given the differences that concern both the share of basic foodstuffs in household budgets and the ability of different stakeholders to deal with price instability. Furthermore, the question of the credibility of any commitments is raised, insofar as food issues are of major political importance and compliance with restrictions may be highly problematic. This problem regarding restrictive commitments is shared by all parties, and should now be included in the scope of any analyses so as to enable more constructive debate on the potential development of international discipline.

The development of quality standards poses four main research questions.

The first concerns the effects of these non-tariff barriers on international trade and its geography. Not all countries will initially be able to comply with standards, and their application can modify the situation of a country in a market. In addition, the ability of producers to comply with more stringent demands may be highly heterogeneous within the same country. Furthermore, a certain number of countries are tending to withdraw from the global marketplace, deciding that it is too demanding, and are investing in regional activities that are less restrictive, notably in Asia. What might be the consequences of this

fragmentation of international markets? The question of the growing importance of emerging countries in this governance by standards merits particularly
close study.

The second question concerns the effects of these standards on different
types of agriculture. Some authors consider that compliance with standards
involves investment and/or certification costs that small producers will find
difficult to bear, notably in Southern countries (Berdegué et al., 2005; Daviron,
2002). This would tend to accelerate their exclusion from the market. Some
authors have emphasised the fact that health crises have driven the industrialisation of agriculture in the name of providing protection against the risks of
epidemics (the slaughtering of free-range chickens in Vietnam during the avian
flu epidemic is an example) (Figuié and Fournier, 2010). In contrast, other
authors consider that when the implementation of these standards is accompanied by technical advice, purchasing contracts and loans, they provide a
means for inclusion in the market and better remuneration for small producers
(Minten et al., 2009; Swinnen, 2007) or farm employees (Maertens and Swinnen,
2009). There may be two opportunities to settle this controversy: the improvement of methods to assess these effects, and the development of panel surveys
so that better account can be taken of medium-term effects.

The third question concerns the effects of these standards on the objectives
they are designed to achieve and their efficiency as instruments for sustainable
development. The specifications for these standards include more questions on
the conditions for their implementation than on their efficiency. Impact assessments still remain too focused on evaluating the efficiency of standards: 'do
standards do what they say they will do?', with little consideration of their
relevance with respect to more global impacts: the social effects of environmental standards, the carry-over effects on activities that are not directly targeted, opposite effects, etc.

Finally, the fourth question concerns the political consequences of the types
of governance associated with implementing such standards. Depending on the
standard, these governance methods seem to differ on certain points (the role of
the State is still too great with respect to health standards, and not great enough
for environmental or social standards) and converge on others; they are now
drawn up by multi-stakeholder forums involving private enterprises, NGOs,
research bodies and sometimes governments. Such partnerships are intended
to involve all stakeholders in order to prevent criticism, but in some cases this
approach will stultify debate on political options and thus concentrate on the
feasibility of instruments. In political science, governance by standards thus
appears to be a fertile field for research.

8.6   Conclusion

By focusing on the international dimension of issues concerning food sustainability, this chapter underlines the importance of understanding the origins and consequences of the volatility of agricultural prices, and how public policies can respond to these situations. It is particularly important to take account of potential effects with respect to malnutrition, hence the need to identify the most vulnerable populations and optimise the targeting and implementation of public intervention when such decisions are made. However, our views are not as alarmist as those of others (for example, UK Foresight; UK Government, 2011) regarding our entry into an era of increased agricultural price volatility. We do not confirm or deny this view. We prefer to highlight the need for international collaboration on trade policies, enabling an optimum response to inevitable sources of uncertainty and hazard.

This chapter also emphasises the roles of norms and standards, the increasing importance of which presages different governance modes linked to the issues of sustainable development. The implications of these different measures, both public and private, need to be further analysed, both in order to understand the conditions required to ensure their efficiency and to clarify their potential effects.

9

Elements for a foresight debate on food sustainability

Authors: TÉVÉCIA RONZON, SANDRINE PAILLARD
AND PHILIPPE CHEMINEAU

Contributors: CHRISTINE AUBRY, NICOLAS BRICAS, PAUL
COLONNA, CATHERINE ESNOUF, STÉPHANE FOURNIER, HERVÉ
GUYOMARD, PASCALE HÉBEL, JEAN HIRSCHLER, CLAUDINE
JOLY, CÉLINE LAISNEY, BERNARD MAIRE, JEAN-LOUIS
LAMBERT, SOPHIE LE PERCHEC, CHARLES PERNIN, JEAN-LUC
PUJOL, BARBARA REDLINGSHÖFER, VINCENT RÉQUILLART,
MARIE RUSSEL AND BRUNO VINDEL

Regional and global food systems are constantly evolving, thus the contextual elements presented in Chapter 1 are likely to evolve, and food systems will be transformed. Because it is impossible to predict the food systems of tomorrow, we have adopted a foresight approach in order to try and understand possible future changes. Our approach, which is presented in the first part of this chapter, has therefore mainly been based on identifying the main drivers of the transformation of food systems. This work was the fruit of collective discussions by a multidisciplinary group made up of some 15 experts. The plurality of their views and their areas of competence allowed them to analyse the potential impacts of the different evolutions identified relative to the sustainability of food systems in terms of their nutritional, economic, social, cultural, environmental and territorial dimensions. This step of the analysis also enabled us to highlight a certain number of points at issue, which are presented in the second part of the chapter; this does not end with a presentation of different scenarios (as might have been expected), but concludes with the

Food System Sustainability: Insights from duALIne, eds. Catherine Esnouf, Marie Russel and
Nicolas Bricas. Published by Cambridge University Press. © Cambridge University Press 2013.

three transversal messages arising from debate by this workshop: issues linked to inequalities of access to food, territorial dynamics and the governance of food systems.

9.1   Food systems evolving under the effects of various factors

Through the identification of factors underlying the transformation of food systems, it appears clearly that some trends have already been identified (see, in particular, Chapter 1 on the context and the challenges of food systems, and the retrospective analysis described in Chapter 2). Nevertheless, these trends involve a certain number of questions and uncertainties, notably regarding the nature and degree of their potential effects. These uncertainties thus open the way to contrasting scenarios for food systems throughout the world (see Figure 9.1).

9.1.1   Demographic dynamics and the organisation of food supply chains

Demographic growth is one of the main trends observed, and suggests that the global population will stabilise at between 8 and 10 billion people in around 2050 (9 billion according to the median prediction by the UN, with more rapid growth in developing and emerging countries; United Nations, 2008). These dynamics raise questions as to the ability of global production systems to assure food security in quantitative and qualitative terms for 9 billion people, a fundamental element in the sustainability of the global food system.

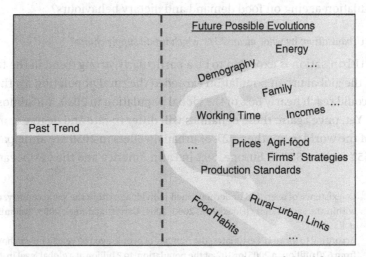

Figure 9.1 Numerous factors governing the evolution of global food systems.

However, more directly, the configuration of food systems will largely depend on how this future global population will be distributed between developing and wealthy countries, between coastal and inland regions, between urban and rural areas. In other words, uncertainties regarding migratory flows (of climatic origin or not) mean that the future spatial distribution of different actors in food systems (consumers, producers, processors, distributors, etc.) remains unclear, and hence the resulting territorial and supply chain organisations: what distances (in geographical terms and regarding the number of intermediaries) will there be between producers and consumers? For what sizes of production areas? For which flows?

In addition, it is difficult to anticipate the geopolitical and economic contexts in which these demographic changes will occur. This adds elements of uncertainty to the evolution of poverty and inequalities worldwide, whether this is in countries that are currently wealthy or poor. And poverty inequalities concern hypotheses on future food production methods (persistence of manual farming methods, development of urban agriculture, etc.), purchasing behaviours (under more or less stringent budgetary constraints), distribution channels (new developments of informal channels), public health (a double dietary burden,[1] the coverage of health-related costs) and regulations (social policies, security nets, etc.).

Finally, if we anticipate today that demographic dynamics will markedly modify the age pyramid, we shall be confronted by an ageing of the global population,[2] that will be more or less acute depending on different countries and local contexts (linked to the emancipation of women, the exodus of young people, urbanisation modes, mortality rates, etc.). But what will be the effects of this population ageing on food demand and dietary behaviours?

9.1.2   *Urbanisation: by what means? For which urban supply chains?*

Urbanisation is expected to be a particularly strong trend in the future. In 2009, the global urban population exceeded the rural population for the first time; it could reach nearly 60% of the global population in 2030, and almost 70% in 2050. Yet, once again, these dynamics will differ in rate and degree from one region of the world to another: 62% of urban dwellers in 2050 are anticipated in Africa, 65% in Asia, 85% in Europe, 89% in Latin America and the Caribbean, and

---

[1] Co-existence of over- and undernourished individuals within the same country, or even within the same family (Doak *et al.*, 2000, 2004; Garrett and Ruel, 2003; Tanumihardjo *et al.*, 2007).

[2] According to mean UN estimates, the number of people aged 60 years or over should rise from 670 million in 2005 (or 10% of the population) to 2 billion at a global level in 2050 (or nearly 22% of the population) (United Nations, 2006).

90% in North America (United Nations, 2010). One might question the factors that will slow down or accelerate these trends, since they could affect the sustainability of food systems and, more generally, the development pathways (distribution of uses for water or land use, urban misery and social tensions, pollution, etc.).

Whatever happens, these statistical forecasts suggest the continuation of the urbanisation of dietary behaviours, characterised by a reduction in the time spent on purchasing, preparing and consuming foods, and leading to increased consumption of processed, prepared and ready-to-eat foods purchased from major retailers, catering outlets or street sellers, or delivered to the home (CIHEAM, 2008; Padilla and Abis, 2007; Stamoulis *et al.*, 2004). Because of greater sedentarity and modifications to dietary behaviours, urbanisation has been accompanied by a very marked growth in diet-related non-transmissible diseases and their associated public health costs (Popkin, 1999; Schmidhuber, 2004; Tsolekile, 2007). The increase in the number of overweight and obese individuals is reaching such proportions that, according to Butland *et al.* (2008), at least 30 years would be necessary to reverse this trend.[3] The urbanisation of dietary behaviours also supposes an orientation of food systems towards massive flows of seasonal and non-seasonal products, towards an enlargement of the sources of supply of these different products and towards the development of public and/or private specifications and quality standards targeting processors and producers, which would increase the costs of access to urban markets for the latter (Reardon and Gulati, 2008; Reardon *et al.*, 2007) (see the analysis proposed in Chapter 6).

However, depending on the different modes of future urbanisation, food systems may present contrasting characteristics. Thus the growth of coastal towns would favour supplies from the sea, with a resulting risk of competition for inland producers. Another possibility might be the urbanisation of secondary towns spread throughout territories, rather than urbanisation concentrated in the principal metropolises of the world. Such an evolution would be likely to promote links between town and country, and hence opportunities for small producers to obtain access to urban markets (Bricas *et al.*, 2009; Satterthwaite *et al.*, 2010; Tacoli, 2003, 2004). This would contribute to slowing the trends towards urbanisation and the homogenisation of dietary behaviours. Similarly, the governance structure adopted by a town will affect its own food

---

[3] In 2005, 1.3 billion adults were overweight and 400 million were obese, 200 million of them in developing countries (Kelly *et al.*, 2008) where obesity and chronic undernutrition co-exist. The authors characterised this trend reversal by a return to the obesity rate that prevailed 30 years ago, together with early signs of a reduction in obesity-related diseases.

system (see Chapter 6 on this subject), depending on whether it devotes resources to securing supplies (storage, green belt), providing security nets against the development of famine (preferential supply chains, grants for food, etc.) or to sanitary control methods for food sectors, or if it does or does not include zones for urban agriculture in its development plans (see, in particular, the World Future Council and Göpel, 2009). The many options will exert a direct effect on the sustainability of food systems. These considerations plead in favour of encompassing current phenomena at several scales. If the global trend is indeed towards an urbanisation of behaviours, how could local dynamics encourage the emergence of new supply models that will respond to local political, social and environmental challenges? This issue has been addressed in Chapter 6.

9.1.3   *New opportunities and constraints on production. What adaptation strategies will be adopted by producers?*

It is generally accepted that the constraints weighing upon food production are likely to increase in the future. Climate change may induce a shifting of production regions and accentuate or create situations of dependence on imported foods, notably and simply because of more frequent extreme climatic events. Large-scale land acquisition could further exacerbate pressure on land use. The need for non-food agricultural products (biofuels, green chemistry) will cause competition between the food and non-food uses of crops and products. Food health and quality standards could tend to develop. The volatility of agricultural and energy prices may accentuate the risks associated with agricultural investments, and feed on the speculative behaviour it has thus triggered. Finally, production systems will have to adapt to a context of competition for uses, and the over-exploitation or qualitative or quantitative exhaustion of natural resources (water, land, fossil carbon, biodiversity). Furthermore, because the gains in labour productivity (in both agriculture and industry) achieved in recent decades were associated with a growing incorporation of fossil energy in production processes, there is a risk that the future dwindling of fossil energy resources may (also) cause a reduction in labour productivity (J.-M. Jancovici, personal communication, 2011).

Once again, anticipating these different constraints will contribute to prospective thinking insofar as the adaptive strategies adopted by agricultural producers may differ (Adger *et al.*, 2005; Bebbington, 1999; Scoones, 2009) and allow us to imagine new configurations for food systems. Will the sum of all these constraints promote the emergence of a new agricultural paradigm, based on the resilience of production and supply systems? Thus 'modern' strategies are currently developing in order to limit risks, such as recourse to insurance-based regimes or the contractualisation of producers with an integrator or fair trade purchaser at a predetermined price. Each of these strategies may cause a new distribution of decision-making between actors in the sector concerned, or even of added value. In this respect, they may induce important reconfigurations of food systems. In a context of greater production system resilience, agro-ecology, which has recently been placed on the political agenda (see De Schutter, 2010), is developing on some farms and is a research priority for CIRAD[4] and INRA (INRA, 2010). Furthermore, with the democratisation of access to information, recent decades have seen the emergence of a new image of modern producers in both Northern and Southern countries: those who are 'connected' to agricultural alert systems and stock markets thanks to information and communication technologies; or producers, who although isolated from agricultural advice services, can now train and inform themselves if they have access to a means of communication in order to optimise their production systems. Should the dissemination of information and communication technologies be seen as a weak signal of an in-depth reconfiguration of agricultural sectors, or just as a simple tool at the service of classic adaptation strategies?

Classic strategies range from greater autonomy of production systems (reduction of external inputs, integration in a short chain), to contracting with an integrator, including the diversification of non-agricultural income streams or even other options. Peasant farmers throughout the world have already proved their capacity for adaptation in the past (Adger, 2003), but we should point out today, in addition to the climatic context, that these farmers will be largely dependent on the prevailing socioeconomic and technical contexts, i.e. the level of agricultural investment, and on their opportunities for access to production resources and techniques and for training. Put more simply, they will to a great extent be dependent on the agricultural model that develops in the future.

9.1.4 Evolution of food firm strategies: integration of consumer concerns for sustainability?

The future orientation of the agri-food system will largely depend on the strategies adopted by the agri-food industry, and particularly by their dominant actors, multinational food firms. The main trend today is towards a concentration and financiarisation of firms, the tertiarisation of activities (R&D, marketing, packaging, transport) and the hyper-segmentation of products. According to Rastoin *et al.* (2010), these orientations extrapolated to 2050 will lead to an in-depth reorganisation of the employment structure, because by that point there will be no more than '500 000 agribusiness firms, each owning

---

[4] http://www.cirad.fr/qui-sommes-nous/notre-strategie/strategie-scientifique.

4000 hectares of land [versus 500 million with 3 hectares today] (...) and about a hundred agri-supply, agri-industry and commercial firms'. They may be able to limit the transmission of price instabilities to consumers, notably via the reformulation of recipes and the replacement of ingredients,[5] and this may be coupled with the acquisition of soft power (advertising and lobbying) by these firms in line with their economic growth. Thus, under their trend scenario, Rastoin and Ghersi (2010) highlighted a risk that firms may take control of the global governance of the food system in 2050 relative to issues such as quality standards for products, information for consumers, tax, etc. The main trend described above targets an increase in labour productivity and capital yield. It could therefore very well move towards a rationalisation of logistics, and hence a reduction in the distances covered (see also Chapter 6). Nevertheless, the sustainability of this trend is widely questioned today. It would indeed participate in eroding domestic biodiversity (FAO and World Bank, 2010) and in greater planetary interdependence, making supply systems vulnerable to contamination by pathogens or pests, whether voluntary or not. The isolation of part of the system following episodes of bad weather or pandemics would indeed affect the system as a whole.

Despite this strong trend, future changes to strategic positioning are conceivable. Indeed, can the current reinforcement of the notion of 'corporate social responsibility' and commitments to sustainable development be perceived as a weak signal of firms taking greater account of social and environmental issues? The increasing scarcity of natural resources may also encourage a greater integration of food and industrial activities linked to biomass, in order to optimise biomass flows via chains that recycle and upgrade by-products (see Chapter 4).

9.1.5   *Reductions in purchasing power: what effects on consumer behaviour?*

Consumers also may increasingly have to integrate new economic constraints in their purchasing behaviours. On the one hand, the price of final food products could follow the upward trend as a result of pressure on agricultural prices (Lambert, 2009) and because of the different pressures on production listed above, coupled with an increase in individual food demands, particularly for animal products. On the other hand, food commodities are increasingly being assimilated with other commodities and are the subject of financial speculation. Their prices could be subject to increasing volatility, whereby major retailers tend to pass on price rises more easily than price reductions

---

[5] 'Concentration in the food industry gives it power for negotiating with suppliers and an ability to adjust to rises in the costs of raw materials through gains in productivity and changes to existing products, at least in SME' (Rastoin and Ghersi, 2010).

(Lambert, 2009). Will these new economic constraints induce significant changes to the purchasing behaviours of poorer and middle-class populations, such as substitutions between animal and plant products, between raw and processed products, between products of different quality or between eating at home or elsewhere? What will be the repercussions of these substitution effects on supply chains (imports) and on retailing systems (increase in hard discount sales)? Could the reformulation of recipes involving the substitution of raw materials have nutritional impacts?

9.1.6   Dietary habits: cultural and psychological obstacles to change, or opportunities for innovation?

Although sociological observations have demonstrated phenomena of 'food resistance' to the homogenisation of dietary behaviours linked to the cultural nature of the diet (Bouly de Lesdain, 2002; Diasio *et al.*, 2009), future decades could nevertheless see profound changes to the representations linked to food.

Could large migratory flows reduce this resistance? By plunging individuals into completely new contexts, migration contributes to a loss of dietary references among displaced populations and a transformation of their dietary styles (Calvo, 1997). It should however be remembered that such transition periods between a traditional and a new diet are often accompanied by nutritional imbalances, particularly in children (Rovillé-Sausse, 1999). At the same time, as a factor for social and cultural diversity, migration often gives rise to culinary innovation in host societies, which little by little adopt new ingredients, new recipes and new cooking techniques (Corbeau, 2000; Régnier, 2006). Such a food 'cross-culture' also implies modifications to existing supply, and even distribution channels (new products, new marketing strategies) (Bouly de Lesdain, 2002; Maire and Mejean, 2008).

Furthermore, the 'acceleration' of social change within generations (Rosa, 2010) may come up against 'food resistance'. Will this acceleration also apply to dietary habits? Will it occur in all parts of the world? If this is the case, a multitude of changes to dietary behaviour could be envisaged until 2050, ranging from the simple recombinations of recipes and dietary practices that exist throughout the world today, to the integration of food substitutes that are currently unknown but are reminiscent of traditional dishes (for example, insect-protein-based meat substitutes), or take on new forms and textures (meal capsules, appetite-suppressant mousses, etc.) and new functions, in the same way as 'nutraceuticals'.

Finally, with an easier information flow and more diversified choices, food is more than ever a focus for the expression of dietary, ethical, moral, cultural and environmental values. Initiated by informed consumers with sufficient purchasing power, this new search for the significance of food products has resulted in a need for new environmental, social and nutritional indicators that are accessible to consumers, and also the creation of new supply chains and distribution channels, new out-of-home dining markets and even new training opportunities (see the Slow Food University in Italy[6]). It has participated in the development of additional market segments and the emergence of new research, development and marketing strategies. According to the common mechanism of values spreading downwards from wealthy classes towards less wealthy classes (see Gojard (2000) in the case of breastfeeding), should we anticipate an evolution in representations of nature, of social relations and of the human body in different social and cultural contexts that will promote the growth of demand for an environmental, ethical and dietetic content in foods throughout the world?

Analysis of this selection of factors for change emphasises the fact that the future will not necessarily be a prolongation of current trends towards a convergence of food consumption, as described in Chapter 2. This trend towards convergence might equally be accelerated, slowed down or thwarted, depending on how the factors for change listed above, or other factors not mentioned here, are expressed. It is indeed highly probable that these factors will be expressed differently according to local cultural, religious, family, economic and geopolitical contexts.

9.2   Questions of sustainability according to food-eater categories

In order to structure an analysis of the sustainability of future food systems, the working group chose to highlight five main categories of food eaters[7] worldwide so as to maintain a balance in its debate on the food eaten by individuals with low or high purchasing power and living in wealthy or developing countries (see Table 9.1).[8] For each of these five categories of eaters, the working group focused on selecting the factors for change that would exert the most influence on the sustainability of food systems (cf. Box 9.1). The discussions considered a timeframe of several decades, without fixing a specific time horizon, insofar as the effects of each of these factors for change will not all comply with the same timescale.

---

[6] University of Gastronomic Sciences, http://www.unisg.it/welcome_eng.lasso.

[7] We prefer to use the term 'eater' rather than 'consumer' in the economic sense, so as to include people eating the food they produce themselves.

[8] The simplicity of this classification induces a certain number of limitations, the main ones being (1) the grouping of consumers in emerging countries with those in developing countries, and (2) the approach per eater rather than per household, thus masking the specific roles of the people designing or cooking the meals.

Table 9.1 *Relative importance of different categories of food eaters throughout the world.*

	Poor food eaters	Wealthy food eaters
Developing countries	Rural food eaters • 2 billion people, most of them agricultural producers[a] • 800 million are undernourished (500 million peasant farmers, 100 million livestock farmers and 200 million who are landless). This global population has been decreasing during the past 40 years, but is rising in sub-Saharan Africa[b] Urban food eaters • 750 million people,[a] including 200 million who are under nourished[b] • 1 billion urban dwellers living in slums, i.e. 43% of the urban population of developing countries (72% in sub-Saharan Africa, 32% in Latin America, 28–36% in Asia)[c]. There could be 2 billion in 2030[c]	Mainly urban food eaters • 46% of the world middle classes (28% in Asia-Pacific, 10% in Latin America, 6% in North Africa–Middle East and 2% in sub-Saharan Africa)[d] • + 1.2 billion people in the middle classes between 1990 and 2005 (80% in Asia, 50% in China), even if the majority only earn $2–13/day[e]
Industrialised countries	• 1.7% (Denmark) to 11.4% (USA) of people in severe poverty[f] • 16.6% of people at risk of poverty in the EU-27 (growing since 2005–08, stagnating in France[g])[h] • Increase in income inequalities (except in Spain, France and Greece) because of the rise of the highest incomes[i]	• Few figures on wealthy populations in industrialised countries • Rise in income for the richest people and fall in income for the middle classes

[a] At the poverty threshold of $2.15/day (Ravallion *et al.*, 2007).

[b] FAO (2006); the most recent figures in this study date from 2003.

[c] United Nations (2003).

[d] Kharas (2010).

[e] Ravallion (2010).

[f] At the poverty threshold of 40% of median income. OECD. StatExtracts.

[g] In France, 4.2 million people were poor, at 50% of the median standard of living, in 2008 (i.e. 7.1% of the population). While the incidence of poverty had been falling since 1970, it started to rise again between 2002 and 2008. Single-parent families (27%), people living alone (16%) and large families (15%) are the most affected by a poor standard of living (ONPES, 2010).

[h] Atkinson and Marlier (2010).

[i] 'The income gap between the richest 10% and the poorest 10% grew by 15% on average [in OECD countries]. Other, more sophisticated, measures of income inequality were 7–8% higher in the mid-2000s than they were in the mid-1980s. This may not sound much of an increase, but it is equivalent on average to taking $880 away from the poorest 50% and giving $880 to the richest 50%, although incomes at every level grew over the two decades.' (OECD, 2008b). http://www.oecd.org/dataoecd/48/56/41494435.pdf).

Box 9.1   Principal factors for the evolution of consumption modes and food systems among the five major categories of food eaters

Category 1 (developing countries, rural poor)

(a) Capacity for resilience and innovation faced with factors for food insecurity (climate change, massive land acquisitions, price volatility, rural exodus, etc.).

(b) Market access, degree of monetisation (share of consumed products that are purchased).

Category 2 (developing countries, urban poor)

(c) Urban agriculture.

(d) Public policies (international trade policies, urban policies and different security nets) and community or charitable actions to ensure food security for these populations.

Category 3 (developing countries, urban rich)

(e) Evolution of consumer practices linked to the development of major retailers.

Category 4 (industrialised countries, poor)

(f) Public policies and community or charitable actions to ensure food security for these populations (same as Category 2).

(g) Transformation of supply strategies in this consumer category faced with an increase in the relative cost of food (notably relative to losses).

(h) Cross-cultural changes, evolution of the diets of immigrant populations.

Category 5 (industrialised countries, rich)

(i) Evolution of life styles and its effects on diet and purchasing practices, including evolutions such as 'tailor-made diets'.

We have only mentioned here the critical points regarding sustainability and the research questions raised by the working group (in italics in the paragraphs below).

The titles of the paragraphs below do not use the exact terminology proposed in Box 9.1: this clearly reflects the evolutions seen as group meetings progressed between the step of identifying the principal factors for change and the formulation of research questions linked to the analysis of the potential consequences of these factors in terms of sustainability.

9.2.1 *What are the determinants of the resilience and innovation capacity of poor rural food eaters–producers in developing countries faced with food insecurity?*

The major determinants of the capacity for resilience of rural food eaters–producers include:

- Individual entrepreneurial qualities and collective capacities for self-organisation by local actors. *How can these be encouraged? What type of institutional support should be provided? What is the role of education in building these skills? How can social innovation be encouraged in this direction?*
- Management methods and access to land and natural resources. To overcome the dichotomy between individual and collective/community property, *what alternative collective regulatory systems should be developed? At what scales? How can customary rights be formalised? And for which development model?*
- The quality of services and infrastructures in rural areas. *What are the investment priorities in these areas (improved access and commercial trade/local supply chains and self-consumption/storage of inputs and foods produced locally and/or imported)?*
- The technical innovation and investment capabilities of entrepreneurs (including the development of appropriate food preservation technologies). *How can the emergence and dissemination of local innovations and/or technology transfer be encouraged? How should access to capital be favoured for those excluded from the banking system?*

If sustainability issues are subject to threshold effects, it is important to better understand *the determinants and thresholds of population density that act on the risks of social conflicts, zoonoses and pollution spread by effluents and waste, by analysing their relationships with specific spatial organisations* (notably, model for the multifunctionality of spaces versus model for the functional specialisation) of spaces.

Finally, if the local governance capacity of a rural area, and its capacity for change, are linked to the dynamism of its rural fabric, this fabric will need to be better characterised. *What are the different components of this fabric (family, neighbourhood, village, union, etc.)? What are the relationships between these different components? Who holds political power? Economic power? Who provides protection in the event of a crisis? What resilience characterises these relationships? How are these relationships likely to evolve in the future (breakdown and recomposition)?*

9.2.2   *What are the effects on local dynamics of greater access for rural*
        food eaters–producers to urban markets?

Changes to the degree of market access for rural food eaters–producers in developing countries will be closely linked to future types of urbanisation (diffuse urbanisation versus concentrated urbanisation), and to the economic and social links between town and country. The corollary of greater access to urban markets will be a trend towards the monetisation of food products (where products that were previously not monetised tend to be replaced by commercial products) and the relative urbanisation of food behaviours, generally associated with a diversification of diets (access to a greater number of products) and also to problems of nutritional balance under these new diets (sugars and fats hidden in processed products, the relinquishment of foods from traditional farming and/ or hunting and gathering).

Entry into the market is almost inevitably linked to social differentiation within agrarian communities. This brings us back to the problem of dealing with inequalities in development models. *Is a development model focused on reducing inequalities profitable to society as a whole, to its cultural progress and to taking account of medium- and long-term constraints (e.g. environment)? Which model is the most efficient in reducing monetary and food poverty: the pursuit of global growth or efforts to reduce inequalities?*

Finally, rural areas in developing countries are now a focus of large-scale agricultural investment. A global governance system that would control these investments through good practice guidelines, so as to reserve part of the added value for local production and/or consumption, is currently under debate (Cotula *et al.*, 2009). *Which political, diplomatic or economic instruments would be most able to ensure the efficiency of applying these standards? Is the importance of brand image sufficient for companies to actually commit themselves to compliance with these standards?*

9.2.3   *What might be the effects of the development of urban*
        and periurban agriculture in developing countries?

The development of urban and periurban agriculture is sometimes put forward as a factor for the sustainability of urban food systems because it may encourage a diversification of diets for the poorest urban dwellers and ensure their food self-sufficiency (see also Chapter 6). Integrated in a production/processing/retail chain in urban areas, urban agriculture may provide both formal and informal employment and act as a factor for social inclusion. Nevertheless, stakeholders in this type of agriculture remain subject to strong competition for land. Furthermore, major questions remain as to the integration of these agricultural systems in the urban fabric: *what risks are associated with the proximity of* these systems to urban populations in terms of exposure to xenobiotics, heavy metals and pathogen micro-organisms, or any other substance used during production? What are the pandemic risks? How can these risks be limited? Whatever the case, any quantitative evaluations of urban and periurban agriculture remain very piecemeal at present (lack of data on the number of households involved, the income accrued, the share in the food market, the share of food consumption by urban dwellers, etc.), and its contribution to development still needs to be clarified. *Does urban and periurban agriculture participate significantly in the income and diet of the poorest people? In which economic and urban contexts?*

Similarly, little information is available on the supply chains and markets for the products of urban and periurban agriculture. *Do these products compete with those from rural areas? With those on the international market? Or do they reach different market segments?*

In terms of cultivated biodiversity, *does urban or periurban agriculture produce the same species and varieties as those available on the standard food market, or are they a source of species and variety diversification that make up urban diets in general? Do they participate in safeguarding domestic varieties that would otherwise be lost?*

9.2.4   *Which drivers could lead to a transition towards more sustainable food systems?*

If the industrial model (specialisation and major retailers) is in the lead today in a certain number of developed and emerging countries, this is partly because it is economically efficient, it has rendered the services expected by consumers and it has directly or indirectly generated new demands from these consumers. *What might be the determinants and mechanisms that would characterise a transition towards new, more sustainable models?*

- The public policies that could steer food systems are very varied in nature (food, environmental, commercial policies, etc.). They also act at different scales: local (e.g. stimulation of local markets by the municipality of Belo Horizonte, Brazil) or international (e.g. establishment of international standards). *Which mechanisms would be the most efficient in coordinating these numerous sectoral policies at different scales of conception and application (cf. Chapter 2)?*
- Consumer awareness also raises *the question of the choice of indicators upon which communication can be based, and notably the methods used to construct these indicators* (cf. Chapters 2 and 10).
- Voluntary commitments by companies constitute a third driver, but *might they not in some cases simply delay the application of restrictive public policies?* As for questions linked to the relative roles of private and public standards, please refer to Chapter 5.

Finally, although food sovereignty tends to overshadow debate on sustainability, the concept itself suffers from imprecision. *Should this sovereignty be envisaged at an individual level (the consumer reigns) or at a more collective level (the nation reigns)? Might respect for the food choices (from production methods to distribution modes) made by a nation not imply non-sustainable effects for other nations? To what extent is food dependence on international markets, and thus on exporting countries, a factor for the vulnerability of food sovereignty in the dependent country?*

9.2.5   *Can the sustainability of long industrial supply chains and smallholder supply chains be compared/contrasted?*

It is probable that smallholder and long industrial supply chains will continue to co-exist, interact and evolve jointly in the future (cf. Chapter 4), each with its own advantages and weaknesses regarding the three crucial points of their sustainability, which are the jobs created, control of their safety quality and the regularity of supplies. At a safety level, the risks of accidents, at the current level of technological progress, are much lower when supply chains are traceable and easily controllable. *Therefore, what conditions should be ensured so that smallholder supply chains can guarantee adequate and controlled safety quality at a lower cost?* Nevertheless, the volumes concerned and the rate of dissemination of safety abnormalities are much greater when the chain is centralised. The same applies for breakdowns in supply: *if local supply chains are vulnerable to breakdown in local production, for example in the event of a climate event, does the fundamental vulnerability of centralised production not reside in a blockage in the mobility of men and goods that might have repercussions at a major scale?*

In industrialised countries, out-of-home dining is now marked by growing industrialisation (school canteens with the centralised preparation of meals, restaurants supplied with processed and prepared products, etc.), and increasingly contractualised supply systems. *Is it possible to reconcile local supplies and industrial catering, and using which organisational instruments for production and logistics? Are small production units capable of complying with the specifications of industrial catering, particularly in terms of quantity?*

The environmental sustainability of different types of supply chain is a controversial subject (see Chapter 6). Indeed, it is now difficult to make an environmental comparison between short and long supply chains or between smallholder and industrial producers. It would be interesting to make such an assessment on Japanese *teikei* and *sankocho*, direct contractual chains between producers and consumers where sufficient experience has now been obtained. Furthermore, depending on the type of product considered, the mode of production and processing of a product may have more effect on its energy balance than its marketing chain (Brenton *et al.*, 2009; Mila i Canals *et al.*, 2007;

Müller-Lindenlauf *et al.*, 2010; Schlich and Fleissner, 2005). This calls into question the often supposed correlation between food miles and carbon footprints. Nevertheless, commercial logistics offer considerable potential for a reduction in the carbon footprint of food products (regional platforms versus multiple points of sale). *What then could be an environmentally sustainable distribution system in developing countries? How would it evolve over time?*

9.2.6   *To what extent are the economic crisis and a rise in the relative cost of food likely to affect food access for poor consumers and the sustainability of food systems?*

A reduction in the purchasing power of poor consumers tends to have a direct effect on their purchasing behaviour, leading to a choice of basic and inexpensive products, which often have a high calorie count (Le Nechet *et al.*, 2006). It is thus correlated with a deterioration in the health of this population via the growth of diet-related non-transmissible diseases. *Will better dissemination of culinary knowledge and skills make it possible to avoid these imbalances? What are the current channels for the acquisition and transmission of these skills? What is the relative importance of factors such as family size or free time when encouraging individuals to prepare their own food? What role is played by women in the choices of the products consumed, and in the transmission of culinary skills and household skills in general?*

In addition, disruption to the working hours of vulnerable workers, which is particularly pronounced in times of crisis, often disturbs the organisation of meals during the day (Millet and Thin, 2005). *What are the consequences of this at the social and individual levels (reduction in time for social contact within and between families, reduction in the conviviality of meals and associated well-being, etc.)?*

We might also anticipate that to minimise their production costs in a context of crisis, food firms would start to relax their qualitative constraints (private standards on composition and organoleptic and nutritional quality, or even in extreme cases the safety quality of products in particular). *What might be the consequences if they did reduce the organoleptic, nutritional and even – in extreme circumstances – the safety quality of the products they sell?*

If access to food for the poorest people guarantees a certain degree of social peace, *might the ability of the food industry to produce foods with a high nutritional value that could be sold at a very low price give them much greater lobbying power, notably with respect to the public regulations that govern their activities (including standards and safety regulations)?*

Still regarding access to food, *are there populations for whom this access is blocked because of a lack of mobility (the elderly living in rural areas, for example), creating the 'food deserts described in the USA? If this is the case, which supply channels do they use? What are the impacts on their diets? How might these problems be prevented?*

In terms of instruments to support food, insofar as the largest budget for underprivileged urban populations in industrialised countries is accommodation, *what would be the effect of public policies aimed at releasing this constraint (rent levels) on dietary practices? Under which conditions might the money saved be allocated to better quality foods or to other types of expenditure?*

9.2.7   *To what extent will migratory flows from developing to wealthy countries induce transformations in the food systems of Northern countries?*

Migratory flows from developing countries to wealthy countries often swell the ranks of poor food consumers in industrialised countries. Indeed, among most of these migrants, the same characteristics as in poor non-migrants are found, and in particular a high incidence of overweight, obesity and diet-related diseases. Children and infants are particularly vulnerable during post-migration periods because they suffer from their parents' lack of understanding of the products available in the host country, and therefore are sometimes given a diet that can lead to deficiencies and malnutrition.

Migration drives the development of new, formal or informal supply chains. Although some data are available in the literature on these sectors (parcels, specialised grocery stores), *there do not seem to have been any studies to evaluate the amount of food imported, by country of origin, as a proportion of the food spend of migrant families. It would also be interesting to determine whether the internet constitutes a new channel for the import of 'ethnic' foods, and whether this is likely to develop significantly in the future.*

In addition, *is the relative price of exotic products higher than that of classic products when the cost of travelling to purchase them is included? In this respect, might it constitute a factor for cultural integration, notably for the many migrants with low purchasing power?*

For the host country, migration provides new economic opportunities in the food sector, whether through the creation of restaurants, street stalls or the emergence of new specialised chains (for the supply-processing-distribution of Asian products, for example). The professionalisation of import activities often gives rise to a standardisation of practices. *Because the introduction of standards and safety constraints causes the standardisation of products, are they a limiting factor on the variety of foodstuffs available? Do they constitute an obstacle to the cross-cultural developments and culinary innovations induced by migrations? Do they result in an increase in the amount of processing, packaging and losses or, on the contrary, greater safety of the product while respecting its original characteristics?*

9.2.8   *Are changes to the lifestyles and social norms of wealthy eaters likely to impact the sustainability of food systems?*

The specific characteristics of wealthier classes reside mainly in their ability to make choices without any strong budgetary constraints. They thus become free, autonomous and reflective food consumers, to use the terms employed by Ascher (2005). They can choose the content of their diet, their supply chains and sites of consumption, as well as the nutritional, organoleptic, environmental or ethical quality of products, etc. all at once, aided in this by different types of indicators. However, *does the availability of information guarantee the spread of virtuous behaviours?* In the case of the USA, it seems rather that it is the problem of overweight people that has led to increased availability of nutritional information. In the same vein, school meals are now strictly governed by nutritional guidelines. However, we have insufficient knowledge on the 'strategies' developed by the users of canteens (exchanges between friends, for example) to know whether these nutritional guidelines actually result in better balanced meals. In France, environmental information has been displayed on packaging or on companies' websites from July 2011, in an experiment which will allow us to refine our knowledge of consumer reactions to environmental arguments. Finally, supposing that the environmental virtues of products may indeed be valued by consumers, this does not exclude *greenwashing* (green marketing) efforts for the sole purpose of creating value for the food industry and/or distribution companies.

At a social level, any segmentation of the market based on a range of more expensive products (e.g. certification) excludes their access for the poorest consumers and questions the democratic nature of the governance of food systems by sovereign consumers. From a qualitative point of view, *might the trend towards certification have the perverse effect of increasing niche markets and only enabling an improvement in the quality of a small volume of the products markets (and not for the majority of consumers)? Is compliance with a specification sufficient to solve the environmental, social or economic problem for which it was initially designed (e.g. FSC label – deforestation, or MSC – seafood, see Jacquet et al., 2010)? Or does it only serve to enable greater transparency? Should an ex-post evaluation of the efficiency of a label be included at the outset in the specification?*

Furthermore, a broader understanding of the impacts of different types of foods on human physiology in a context of 'worship' of the human body has led to the emergence of orthorexic behaviours.[9] The drawback of a trend towards orthorexia resides in the risk of losing the pleasures associated with food: the pleasure of indulgence, the pleasure of time spent together, etc. This type of behaviour might be reinforced by future technological advances. For example, mobile phone applications able to detect the presence of undesirable products could be further developed. It is nevertheless difficult to anticipate the reactions

---

[9] Orthorexia (from the Greek: orthos, correct, and orexis, appetite) is considered as an eating disorder characterised by a fixation on ingesting healthy food. It is related to anorexia and bulimia.

of consumers to the development of information technologies in the field of food. Again at the technological level, *can we expect the introduction of products that 'correct' the harm done by an unbalanced diet? Might they not then involve a risk of losing the social and individual regulation of dietary behaviours, while at the same time encouraging excessive behaviours?*

It is also possible to imagine that technological innovations may render the establishment of individual food profiles a routine matter, at the incentive of public authorities or private stakeholders such as insurance companies. This will not be without posing certain ethical questions: *at which stage in life should such a profile be compiled? At which time should the risk factors detected be explained to the patient? Will such a diagnosis be covered by professional secrecy or should it be communicated to the family, a future spouse or the public authorities?*

In the literature, several foresight scenarios have been built on the development of RFID (Radio Frequency IDentification) microchips (Feillet, 2007; Premier Ministre and Centre d'analyse stratégique, 2009). Demand for these microchips now mainly emanates from producers and distributors, for whom they offer major advantages in their logistics management. *Should we fear that the opportunities they offer to consumers might be insufficiently exploited, even if ultimately it is the latter who could benefit from (and pay for) the services rendered by these technologies (measurement of carbon emissions, detection of allergenic foods, nutritional properties, etc.)?* This point highlights once again the importance of innovation, logistics and industrial organisation to the trends that will affect food systems in the future.

Finally, *will technologies incorporated in foods and packaging consume non-renewable resources? Is it possible to enable the massive development of recyclable or biodegradable packaging systems? If the packaging produced using biomass replaces most other types of packaging (for food and other products), what productive land areas will this represent?*

The comparative analysis of the carbon footprints of different diets, as described in Chapter 3, clearly emphasises the complexity of the nutrition–environment relationship in food.

9.3   Questions for research

The working group identified a certain number of points which led us to highlight three principal and unavoidable issues in our discussions on research relative to food sustainability: (1) the effects of inequalities of access to food on the sustainability of food systems, (2) the links between territorial dynamics and industrial management strategies, and (3) the sustainability of governance modes.

9.3.1   What are the effects on the sustainability of food systems of inequalities of access to a balanced diet (in quantitative and qualitative terms)?

Current trends point to factors liable to deteriorate access to a balanced diet (increase in the relative cost of food, price volatility, problems of access linked to a lack of infrastructure and distance to markets in developing countries, urban *food deserts* in the USA, rural *food deserts* in Europe,[10] climate change, landless peasants, etc.). This weakening of the structure results in adaptive strategies (dietary changes and a search for less expensive products, recovery of unsold goods, urban agriculture, migration/rural exodus, etc.) which are likely to have important effects on the sustainability of food systems, whether in environmental, nutritional or socioeconomic terms. It is important to better understand these strategies (and their impacts in terms of sustainability), which often take the form of informal supply chains, in order to design nutritional policies that are both efficient and more in line with the objectives of sustainable development.

Wealthier households can make their food choices according to their nutritional and environmental concerns, but defence of these 'causes' seems to be out of the reach of households with weak purchasing power (this reasoning also applies to some extent to the North–South dichotomy). The factors mentioned above are likely to further reinforce this divide. The development of more 'inclusive' approaches seems determinant if we hope to take up the challenge of sustainable development.

The co-existence of deficiency malnutrition on the one hand, and an excessive food intake on the other, is becoming common, whether in developing countries (a double burden) or in industrialised countries (reappearance of deficiency malnutrition). Understanding such divergent nutritional trends within a single targeted entity is difficult and complicates the definition of nutritional policies.

9.3.2   How can territorial dynamics and industrial strategies be combined?

The territorial origin of agricultural products and its recognition by consumers who tend to purchase products with a known and certified origin, form a powerful tool for the spatial and human organisation of territories. At the same time, this is one of the essential elements in sharing added value throughout the chain from the producer to the consumer. Inversely, the processing and distribution industries seek to maximise their profits by procuring their supplies at the lowest possible cost but with consistent quality. They also prefer the

---

[10] For example, pauperisation affecting elderly people in rural areas can result in poor mobility, which exacerbates the problem of their access to a balanced diet.

standardisation of raw materials, which enables them to optimise industrial processing and distribution methods, thus prioritising supplies independent of their territorial origins. These supply models, which diverge both in terms of the distance between production and consumption and their logistical organisation, also largely depend on the type of urbanisation in different territories and the links between urban and rural areas.

Neither of these two models will probably win the battle completely, and these two extremes will have to be combined so that they can co-exist and optimise the sustainability of food systems. Therefore, this combination, which concerns complementarities more than an opposition between the two systems, should be the subject of research efforts.

Furthermore, territories may have to manage new divides resulting from the large-scale land purchase/rental. Research will certainly be required to analyse the economic, sociological and other changes that these transactions will cause, and inform initiatives to draw up codes of good conduct at an international level.

9.3.3   Modes of governance for the food system: between the 'over-responsibility' of consumers and the regulation of supply?

Research is participating actively in the development of indicators for the sustainability of products, supply chains and food systems. Research efforts should be pursued, notably in order to better integrate the complexity of systems and the numerous dimensions of sustainability. However, the idea of a complete 'objectivation' of sustainable development leads us to believe that better information of the public (on the nutritional, ethical and environmental qualities of products) would allow us to meet the challenge of sustainable development, by allowing consumers to improve their behaviour. Thus the desire to objectify sustainable development involves a risk of moving it outside the scope of political debate and action.

Better information contributes to greater transparency. However, it could result in more confusion, have anxiogenic effects or even arouse feelings of guilt, by making consumers over-responsible.[11] Under an extreme scenario, it might be accompanied by food companies denying responsibility for the harmful effects associated with their products once they have fulfilled their duty of keeping consumers informed.

---

[11] Consumers whose ability to choose is limited by their purchasing power will be deprived of their ability to claim new values and to drive the food system. See Chapter 2, section 2.5.

9.4   Conclusion

This initial foresight exploration focused on food sustainability aimed to identify the questions that could be addressed by research teams. We have deliberately chosen not to consider the question of food via highly aggregated indicators on production and consumption levels, as has often been the case in recent international foresight studies (Carpenter *et al.*, 2005; CIHEAM, 2008; Foresight Government Office for Science, 2011; McIntyre *et al.*, 2009; Paillard *et al.*, 2010; PNUE, 2007). Unlike global foresight studies, we have also tried to take better account of processing, distribution and catering aspects. Finally, the emphasis laid on the sustainability of food systems has also led us to highlight more qualitative issues linked to inequalities and poverty, and the governance of food systems.

The first lessons that emerge from this work are original, in the sense that they are specific to food and may contribute to the implementation of a more global foresight study that takes account of other determinants and generates new scenarios.

10

A critical panorama of methods used to assess food sustainability

Authors: JOËL AUBIN, CATHERINE DONNARS,
MARKETA SUPKOVA AND BRUNO DORIN

Contributors: GÉRARD GAILLARD, FRÉDÉRICK GARCIA,
LUCIE GAUTHIER-DELTOUR, EMMANUELLE HENRY
ERIC LABOUZE, SOPHIE LE PERCHEC, SYLVAINE LEMEILLEUR,
CATHERINE MACOMBE, THURIANE MAHÉ, JEAN-PIERRE
RENNAUD AND CAMILLE ROJOT

The duALIne project chose to examine the methods used to assess food sustainability in a chapter of its own, separate from the sectorial approaches presented previously, so that this examination could be as open as possible. This chapter focuses in particular on the specific issues posed by food vis-à-vis the methods currently used to measure sustainability. Under this approach, this chapter looks firstly at the complexity of food systems, then how the associated challenges of sustainability could be structured and finally presents some methods and indicators and the research questions they raise.

10.1 Introduction

Measuring performance has become a widespread activity in modern societies. It is the benchmark by which political and economic choices are regularly backed and/or justified. Performance indicators, whatever their objective, have seen exponential development, as have the operators who construct them. Assessing the performance of food systems through the prism of sustainable development is still a recent concern that requires in-depth reflection, both in terms of its scope and of the issue(s) to be assessed on the one hand, and

Food System Sustainability: Insights from duALIne, eds. Catherine Esnouf, Marie Russel and Nicolas Bricas. Published by Cambridge University Press. © Cambridge University Press 2013.

regarding the choices of the sustainable challenges targeted or the assessment methods to be used on the other.

We should first address the intent that underlies any assessment. Depending on the objectives, the methods used will vary. If the aim is to produce a diagnosis, it will be necessary to characterise or qualify the states, functions and dynamics associated with food systems. If the assessment is made in the context of normative approaches, it will verify whether the food system or nutrition style is performing adequately in terms of compliance with standards or norms. An assessment may sometimes implicitly aim to make a prediction or adopt a foresight approach. For example, the aim may be to anticipate the consequences of changes in practices due to external evolutions (climate, safety, etc.), or even to test contrasted or plausible scenarios for the future. The assessment then seeks to demonstrate and describe mechanisms. It may lead to a modelling of phenomena (such as modifications to processes or organisation) or to simulate behaviour (e.g. by consumers). An assessment may then help to determine choices regarding technical or organisational development. It thus supports decisions – notably of a public nature – and may lead to practical applications. For example, the French Environment Round Table (*Grenelle de l'Environnement*) was able to guide regulations in terms of environmental labelling on food products.

The increasing importance of the concept of sustainable development has significantly altered assessment criteria. The Brundtland report (Brundtland, 1987), which initiated the concept, clearly demonstrated that account must be taken of the needs of future generations and any assessment must not be restricted to the present time. Similarly, the spatial scales have broadened: by considering that actions in one place might have a global effect and/or an effect elsewhere, an assessment must take account of local, regional or global processes or phenomena. In the same vein, the context of sustainable development constitutes a challenge of complexity for assessment methods: a product, service or action forms part of a 'system' whose limits, components and their properties must be defined, as well as the internal and external interactions with the specific social, economic, physical and biological contexts. Assessments based on a single time-frame or geographical scale, and on a single indicator, are no longer considered relevant, as acting on one component can generate unexpected parallel or indirect effects. For this reason, preference is now given to multicriteria approaches that aim to qualify food systems in terms of their economic, social, environmental and nutritional characteristics, in conjunction with their spatial and temporal dynamics. These assessment methods are less developed, insofar as combining assessment frameworks remains problematic.

Finally, stakeholder involvement participates in the governance of sustainable development. In the same way as food styles, practices and choices are conditioned by their cultural, social, economic, political and geographical characteristics, assessment approaches now increasingly include different stakeholders in order to jointly build the framework, methods and hierarchy of their indicators.

The scope of any assessment associated with food practices and styles is thus particularly vast. The present chapter will not attempt to cover all the questions associated with assessment methods, but to prepare the ground for a certain number of key questions addressed from the angle of food sustainability. This chapter is based on bibliographical evidence and on the experience of practitioners.

10.2   Complexity of the 'food style' component within an assessment framework

Nutrition style is a notion that lies at the crossroads between physiological needs and social and economic contexts. It covers practices and a combination of different determinants, which are associated with food chains (considered as the succession of stages in product processing) or 'process trees' (combinations) that may be independent or associated together. Depending on whether one is working at the scale of an individual, a community, a sociocultural group, a country or a continent, the determinants of the system will vary.

The simplified representation of the production chain for broiler chickens shown in Figure 10.1 clearly illustrates the complexity of a food that forms part of a food style. In addition to its nutritional functions, this product represents a 'system', the components of which participate in exchanges and/or in phenomena of compensation, substitution or even 'recycling loops'. In addition, this system reacts to its environment: fluctuations in the prices of cereals used to feed livestock, or upstream changes due to land use dynamics, which influence the system as a whole. New characteristics arising from a recombination of these elements may emerge. The performance thus evolves as a result. The challenge facing assessment is thus to take full account of this systemic complexity without breaking it down into its different parts, which would cause it to lose its interaction characteristics.

10.3   Defining the challenges of sustainability before choosing the methods

This illustration of the food chain based on a single food does not however reflect the challenges for sustainable development of social choices

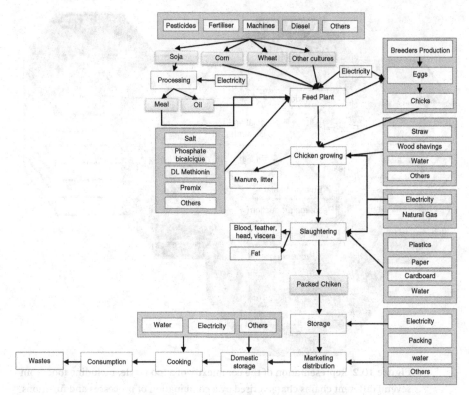

Figure 10.1 Simplified representation of the 'broiler chicken' chain, defined as a system in the context of an LCA of meat chickens. Adapted from da Silva (2010).

in terms of food. However, these choices are determinant in the assessment: which purposes, which priorities and which scales should we be targeting?

For this reason, we felt it was important to adjust our exploration of methods to the challenges defined for food systems. To achieve this, we cross-analysed the functions of food – hedonic, economic, environmental, social, cultural, nutritional, ethical – with the stages or major components in the food system, from the production of primary resources to consumption and the end of life of foods and by-products. It was then possible to question the challenges of sustainability for each combination of functions and stages: conditions for production and trade, food choices, external forces acting on governance of the system (regulations, legal framework), or via markets (marketing, discrimination between products in terms of price, substitution effects, rebound effects, etc.), technological processes, etc.

Figure 10.2 illustrates the construction of this approach: the functions form the different parts of a polygon and the stages constitute the horizontal strata. If the analytical scale is that of foods (represented by a column), the food system

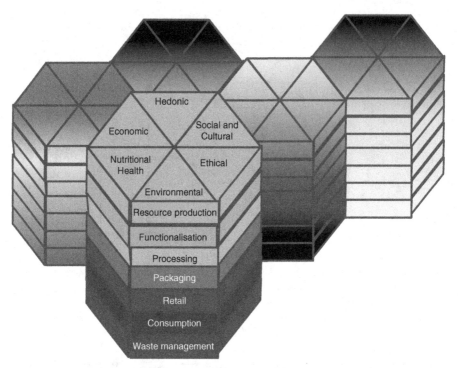

Figure 10.2 Representation of the complexity of a food style, including food from several different chains characterised by a combination of processes and functions.

then corresponds to the combination of the polyhedrons obtained. This representation illustrates the high degree of complexity of the assessment. One could imagine that this scheme might be appropriate for different scales (food for a population, region or nation or the whole world).

It should be noted that the challenges associated with the functions of food are formulated differently depending on regions and cultures. Similarly, the boundaries between stages in the food chain can be segmented differently, or may even 'fluctuate', notably between processing, distribution and associated services (differentiation delayed by the food industry, delivery services associated with distribution, etc.).

This approach allowed us to construct an inventory-style matrix for each combination of processes and functions, questions and/or challenges that could influence the level of sustainability of food functions (Table 10.1). For example, how does meeting the nutrient needs of the whole population modify equilibriums in terms of resource availability?

These challenges illustrate the diversity of inputs and scales at which assessment can be applied (the term 'challenge' taken here in its broadest sense,

Table 10.1 *The diversity of challenges of sustainability associated with the functions and stages of a food system.*

Stage in the food system	Environmental functions	Nutritional and health functions	Socioeconomic functions	Societal functions (cultural, religious, ethical, hedonic)
Production	To optimise the management of natural resources, prevent the destruction of habitats and the extinction of species.	To seek a diversity of food resources.	To ensure that the food behaviours deemed most sustainable also create sustainable employment for primary producers.	To promote an ethical and social value based on product origin that reflects the living conditions of producers.
	To promote production methods that are more environmentally friendly and closer to consumption areas.	To prevent or reduce the use of chemical inputs.	To create synergistic effects in the use and production of resources.	
	To balance the ratio between animal and plant production.	For poor peasant farmers, to encourage production systems that fulfil the nutritional needs of the family; to assess nutritional changes when peasants migrate to towns.	*To achieve the quantity and quality of food products necessary for all populations throughout the world.*	
Processing	To optimise water and energy consumption during processing.	To improve the preservation of nutritional quality throughout the processing chain.	To drive industrial innovation towards greater economic sustainability (sustainable products and processes).	To ensure and qualify the value of industrial innovations.
	To introduce innovative processes.	To limit the use of chemical additives or preservatives during the processing stage.	To encourage social equality.	To promote the social control of innovations.
				To promote 'diversity' and interactions between the functions fulfilled by distribution and/or processing in order to better respond to needs in terms of work, energy, value creation, etc.
				(broadening of 'delayed differentiation' throughout processing).

Table 10.1 (cont.)

Stage in the food system	Environmental functions	Nutritional and health functions	Socioeconomic functions	Societal functions (cultural, religious, ethical, hedonic)
Packaging	To limit the volume of packaging and promote eco-design.	To adapt portion sizes to prevent over-consumption. To improve health protection. To increase the lifespan of foods by preventing recourse to materials that are harmful to human health.	To optimise packaging systems (volumes, functions, recycling). To exploit packaging as a source of information on the social and environmental responsibility of products and suppliers (labelling).	To understand how the perception of packaging influences the value of the contents.
Retail	To optimise food logistics (distribution chains, flow management, sharing of redistribution platforms). To optimise the energy efficiency of retail outlets. To preserve regional biodiversity within the framework of its development (commercial urbanism) and reduce the carbon footprint.	To qualify the global food system with a standardised supply as well as diversified supplies linked to local food traditions.	To assess the economic and social impact of different distribution chains at different scales. To better manage the impact of the installation of hyper- and super-markets (diversity as a source of food system robustness). To understand the roles of different distribution chains in the globalisation of food systems. To encourage potential synergies between flows of materials and services (shops that are not purely centred on food products).	To promote the psychological, philosophical and social determinants (social and economic class criteria) that drive the 'sustainable' purchasing of foods. To encourage 'diversity' or interactions between the functions fulfilled by distribution and/or processing in order to better respond to needs of the population (work, energy, value creation, etc.). *In parallel with the delayed differentiation mentioned with respect to processing stages, but relative to different products and services.*

Consumption	To increase the energy efficiency of final processing at home. To fight against food wastage. To raise consumer awareness as to the environmental impact of food choices. To reduce the environmental impact of travel between the sites of purchase and consumption.	To guarantee a nutritionally balanced diet for different categories of the population. To encourage healthy eating behaviours (menu composition, preparation methods).	To ensure the economic accessibility of food products for all populations. To provide better information for consumers (labelling, basic education). To ensure the viability of diverse food supplies.	To assess the ethical aspect of a product put on the market (in addition to International Labour Organisation (ILO) standards). To assess (quantitatively and qualitatively) the importance of the hedonic function of food. To explore the sources of 'dematerialisation' of eating at home and outside: to what effect (e.g. influence on satiety)? To take account of religious values without encouraging community segregation (costs induced, collective organisation).
Management of end waste	To promote the recycling and re-use of waste. To transform food waste into potential sources of energy (biomass) or to recycle as by-products.	To optimise flows in mega-cities to reduce food wastage to a minimum. To recycle food waste as processed food (without passing through composting and agriculture).	To develop and promote recycling as an economic activity. To integrate the recycling of materials in waste management.	To further explore how the selective sorting of household waste is organised at the individual and collective levels (architecture, urbanism). To encourage a mindset and socioeconomic system that favours a 'second life' for items (recycling, second-hand sale, re-use for other purposes).

Table 10.1 (cont.)

Stage in the food system	Environmental functions	Nutritional and health functions	Socioeconomic functions	Societal functions (cultural, religious, ethical, hedonic)
Transversal processes	*To limit the environmental impacts of food systems.*[a] To reduce dependence on fossil fuels, optimise flows of materials and energy and optimise volumes. To limit pollutant emissions and noise pollution. To ensure better links and cooperation between stages in the system and between different sectors (management of by-products).	To optimise the flows of transported products in order to guarantee their freshness and nutritional composition.	To ensure a balance of skills within the food system to prevent them falling below a socially/ethically acceptable threshold (unqualified labour, etc.). *To spread value throughout the food system.*[a] To distribute decision-making and social skills within the chain. *To create employment* (link between the complexity of a food system, employment needs and the qualifications for these jobs).[a] *To determine how food systems can be adapted in the event of crises or external constraints.*	To improve the skills and know-how associated with food systems (particularly culinary, but not solely). To manage all stages in the food system in a sustainable manner. To gain a clearer understanding of the social impact of risks affecting the sustainability of a food style (transgression of cultural or ethical rules; health risks, e.g. obesity, cancer, toxic risks/ pesticides, fertility, etc.); societal risks that modify food systems (e.g. accept global warming by 2°C).

[a] Issues developed in Table 10.2.

including operational objectives and/or research questions). Many of these challenges correspond to the questions that are found in different international studies. They correspond to some of the 'Top 100 questions of importance to the future of global agriculture' (Pretty *et al.*, 2010), in particular concerning wastage, accessibility and governance. The British foresight study, 'The future of food and farming: challenges and choices for global sustainability' (Foresight, 2011) also highlighted these challenges. Table 10.1 presents some of them as defined by our working group. Our objective was to highlight the relevance of assessment methods in light of the challenges thus defined. We shall be base this work on the challenges proposed in Table 10.1 (in italics), illustrated by the methods in Table 10.2.

10.4   Environmental, social, economic and nutritional methods and indicators

The challenges of sustainability defined by society (arbitration or consensus) will determine the criteria to be qualified using appropriate methods and indicators. It is thus these challenges that will target and define a particular method. Table 10.2 illustrates this approach. There are very many methods of assessing sustainability, but few of them focus specifically on food. One method alone will never fully respond to a complex challenge, and it seems preferable to combine different methods and indicators to meet the objectives of the analysis, although we might not always know how to achieve this. The relevance and limitations of the results of studies and research, like the choice of indicators, can also be assessed by considering the challenges defined. The principal bibliographical references associated with these methods are presented at the end of this chapter.

Most assessment methods generate indicators. According to Bockstaller *et al.* (2008), indicators can be of different orders of magnitude (data, calculations, observations, measurements) and supply information on variables that are more difficult to access, or on more complex systems. These indicators should help users regarding their actions (decision-making, construction of an action programme, modelling). Their use is justified for metrological reasons because of the complexity of measurements (e.g. biodiversity), of the system (food system) or of the concept (sustainability); or for operational reasons, such as excessively expensive measurements (Bockstaller *et al.*, 2010). The OECD (1997) defined the principal types of indicators, the typology of which remains widely applicable: driving forces, pressure, state, impact, response.

The European Commission (2005) proposed estimating the quality of an indicator on the basis of five criteria (using the acronym: *RACER*): it must be

Table 10.2 *Examples of methods and indicators that respond to the challenges or issues of sustainability (the number shown in the 'Methods' column refers to the bibliographical references listed at the end of this chapter).*

Challenge	Sustainability criteria	Methods	Examples of indicators
To generate the quantity and quality of food products necessary for all populations of the world	Availability Accessibility (both temporal and spatial) Quality (gustatory, health, etc.) Equality/fairness Productivity Efficiency	HDI (Human Development Indicator) Mass balance Materials balance Input/output	– Life expectancy – Level of nutritional education – Nutritional quality (calories, protein, diversity, etc.) – Health standards – Incidence of diseases due to over-nutrition, malnutrition and under-nutrition – Yields
To limit the environmental impacts of food systems	Critical natural resources Host capacity and living conditions of a territory Protection of the environment	Life cycle assessment Ecological footprint Water footprint Carbon balance (PAS 2050) Carbon footprint	– Water consumption – Energy consumption – Greenhouse gas emissions – Biodiversity indicators – Ecotoxicity – Land use and quality – Eutrophication – Quantity and quality of waste
To distribute the value chain throughout a food system	Equality/fairness Distribution of profits Wealth creation	Porter Value Chain Life Cycle Costing Contribution to wealth	– Distribution of added value – Market price (absolute or relative to the legal minimum wage)

| To create employment (quantity and/or quality) | Contribution to economic activity
Social integration
Ethics | Productivity (GDP, NDP, etc.)
Social LCA
Global Reporting Initiative (GRI)
Corporate Social Responsibility (CSR, SA8000)
ISO 26000
ILO standards | – Economic network and regional development
– Contribution to the well-being of the population
– Number of jobs produced
– Level of pay
– Freedom of unions
– Equality, accessibility
– Child labour |
| To ensure that a food system can adapt in the event of crises or external constraints | Efficiency
Stability
Reliability
Robustness
Resilience
Adaptability | Cost–benefit analysis
A method crossing LCA and consumption risk assessment (statistics)
Impact studies | – OECD indicators: pressure, state, response
– Productivity of resources |

Note: For further details of methods, see references at the end of this chapter.

Relevant with respect to the objectives, *Accepted* by actors and stakeholders, *Credible* for experts and interpretable without ambiguity; *Easy* to apply, and finally *Robust* towards sources of errors or manipulation.

A few families of emblematic methods and indicators are presented below.

10.4.1   *Life cycle assessments and derived methods*

Life cycle assessments (LCA) currently occupy a central role in assessing the sustainability of food systems. They have the advantage of dealing with different challenges at scales that range from local to global, and they can cover the entire chain from field to fork.

LCA is an environmental analysis method that is now standardised (AFNOR, 2006a, b). It quantifies the environmental impacts of a product or service throughout its life cycle, from the extraction of raw materials to the end of life of products: destruction or recycling. LCA considers all resources used and all pollutants emitted at each stage of the development and destruction of products. It generates indicators of environmental sustainability that can be aggregated at different spatial scales, and does so, relative to a functional unit characterising the objective of the system under study (e.g. kilogramme of product, hectare × year, kilometres travelled, etc.).

This method has numerous advantages. It is recognised, standardised and benefits from dedicated software programs. It can cover the whole of a sector, highlighting the flows of energy and materials that are consumed and emitted. It can also take account of various environmental impacts and, at the present time, is the only method whose scope can be 'cradle to cradle'; in other words, it takes account of the complete recycling of all waste and by-products. It generates new data and clearly highlights the stages (or processes) that contribute to environmental impacts. It is therefore useful to rank different issues (e.g. comparing the impact of transport versus the production stage) or to make a comparison between two processes. In this sense, LCA is a tool that can assist decision-making based on the environmental performance of a product, system or process. It is a tool that can trigger alerts or awareness, being designed as an aid to decision-making rather than a decision-making tool in the strictest sense.

However, these qualities are moderated by several limitations. Classic (so-called attributional) LCA is a static method that does not take account of the dynamic nature of phenomena. LCA is based on a certain number of rules and hypotheses (the frontiers of the systems studied, impact allocation rules concerning by-products) that may sometimes have a significant influence on outcomes (Dorin and Gitz, 2008, in the case of biofuels). The norm recommends measuring the impact of these choices on the results using sensitivity analyses. LCA was designed for assessments at the scale of a product or services. Its extrapolation to the scale of a business (e.g. a farm) has been shown to be particularly useful as a tool for environmental management (Rossier and Gaillard, 2004). Nevertheless, the same approach is not usually applied when a regional or national scale is targeted: the LCA method is relatively difficult to implement for national assessments because of the multitude of hypotheses regarding the delimitation of boundaries, or the rules of aggregation and allo-cation between products (to prevent double counting). In addition, for this type of global study, it is difficult to use databases from different sources because they may take account of rules on impact allocation that are not necessarily compatible with each other. Attempts have nevertheless been made in this respect, such as the 'hybrid' LCA at the scale of a country (Weidema, 2006). In this type of study, the level of impacts is associated with the global yield of the system (efficiency). Furthermore, the issue of losses, and their status within the system, is particularly important: LCA is based on economic criteria, i.e. any-thing that has a value is considered as a product (or by-product) and not as waste.

Little account is taken of the direct and indirect uses of land, changes to habitats and impacts on biodiversity, and their (difficult) assessment varies depending on the studies. Human, ecological and toxicological impacts are often included, but they can vary markedly in the different models employed. Impacts on the working environment and on social organisation are still miss-ing from this assessment. A harmonisation of LCA methods and of data acqui-sition is necessary to improve the comparability of LCA results. A combination of LCA with different scales of food systems would enable the study of the complexity of these systems and generate information on the diversity of challenges. It should however be noted that the limitations of LCA are often generic and also affect other methods.

LCA is the leader in a series of methods that extend beyond the framework of environmental analysis to cover the economic and social domains: *life cycle costing (LCC), total cost accounting (TCA), total cost of ownership (TCO), hybrid analysis, life cycle activity analysis (LCAA), life cycle optimisation (LCO)*. A comparative analysis of these different LCA-based methods was carried out during the CALCAS study (Zamagni *et al.*, 2009).

A method that is still under construction is the '*social LCA*', which links the concept of life cycle assessment to development theories. Because its epistemic foundations are not yet stabilised, there is no consensus as to its inventory indicators among the 200 that have been proposed (REAP, 2010). Kruse *et al.* (2009) made a distinction between quantitative (or additive) indicators such as the costs of production, labour or industrial accidents, and qualitative (or descriptive) indicators which could be the absence of forced labour, freedom of association, compliance with regulations, etc. Although additive indicators can be reflected in the functional unit chosen for LCA, this is not the case for descriptive indicators, thus opening the way to a new field of research. Some studies have focused on the causal relationships (pathways) that link inventory indicators to social objectives (e.g. life expectancy). Weidema (2006) is one of the few authors to have sketched out these pathways to generate reference points. He studied the use of a common assessment unit such as DALY (Disability Adjusted Life Years, an indicator of life expectancy in good health developed by WHO). Nevertheless, much still needs to be done to convert all impacts (both environmental and social) into DALY units, and generally speaking this indicator remains controversial. Finally, studies on domestic food practices by consumers, and recycling, are particularly rare; and indeed, this is one of the strongly anticipated outcomes of social LCA (Jørgensen et al., 2009).

Box 10.1   A critical lack of available data

A method or indicator is chosen according to the initial question to be answered, but this also depends on data availability. Today, these data are insufficiently adapted to the purpose of assessing the sustainability of food systems. This is notably the case of methods such as LCA, which require an analytical approach and complete, accurate information on stages, materials and flows. In this respect the Agri-BALYSE database, which is currently being developed on behalf of the ADEME, should facilitate the environmental assessment of food systems with respect to primary production.

Some data arise from public statistics (in Northern countries) on household purchasing and on the import, sale and export of raw and processed agricultural products. Inventory data are also available on the production and processing of agricultural products, although their accuracy varies in different sectors. This information is even more deficient with respect to Southern countries. Nor are there sufficient data on losses, wastage and waste management (cf. Chapter 7). There is also a lack of post-distribution information on domestic practices. Observational studies can sometimes combine static methods (assessments) with dynamic methods (trends). A harmonisation of protocols for the acquisition of inventory data is essential to facilitate the interpretation and comparison of results between studies. However, the use of mean or constructed data to perform comparisons over time, between sectors, regions of the world, foods or food styles would be hazardous in view of the existence of major differences in context at various levels (ecosystems, technologies, socioeconomic situations, etc.).

10.4.2 Approaches derived from economic sciences

These approaches are useful insofar as they can translate the characteristics of food systems into a cost or benefit for society. The European Commission thus explored the environmental impact of products consumed within the European Union (EIPRO – Environmental Impact of PROducts (phase 1); Tukker *et al.*, 2006) using methods based on *an analysis of inputs and outputs*, sector by sector. These tables are produced by statistical agencies through a description of production and consumption data. The structure of the model consists in quantifying the relationships between production and consumption systems in terms of purchases, sales, resource use and emissions. The model covers the entire life cycle of products. The technological matrix highlights interactions between activities, while the environment matrix quantifies inputs in terms of resource use and outputs in terms of emissions and environmental impact (acidification, ecotoxicity, global change, ozone hole, oxidation, etc.). These input–output tables thus clarify the nature and degree of different environmental impacts, sector by sector.

Although the principles of input–output analysis are quite simple, the analytical data are not always accessible, homogeneous or measurable (financial transactions do not cover all the uses concerned).

These input–output methods can be used at global environmental scales, but not at a detailed level. The European study highlighted the considerable environmental impact of food and drinks.

Some authors have focused on the *cost as an indicator of food efficiency and/or sustainability*. Heller and Keoleian (2000) thus evaluated the cost of food as a percentage of available income or workdays necessary. For example, an average American consumer spends 10.7% of his available income on food, which is equivalent to 40 days of work per family per year (data for 1996). This indicator can also be used to follow trends over time of the share of food in household budgets, and to compare the situation in different countries. For example, the USA is one of the countries where food is the least expensive relative to household income, and India one of the most expensive. Furthermore, Cuellar and Webber (2010) compared the *energy cost of food*: the fossil energy necessary to produce food ranged from 8% to 20% of national energy consumption (studies in Northern countries). These calculations contained quite large margins of error (up to 20%). Nevertheless, these approximations underlined the importance of production (a third) in the chain up to consumption, and showed that dairy and vegetable production are among the highest consumers of energy. Energy wastage due to losses could also be estimated: it was evaluated at 27% of the energy devoted to food and at more than 2% of the total energy consumed in the USA. These approximations are designed to constitute alert thresholds: for example,

Cuellar and Webber deduced that the total energy wasted through food losses was greater than the energy produced in the form of ethanol and inland oil in the USA.

A second series of methods compares the cost or efficiency of public actions or policies. Thus *cost–benefit analysis* reflects impacts that are quantifiable in monetary terms. It can also integrate environmental criteria (CAS, 2009). *Cost–efficiency analysis* also compares impacts between different options. These methods can be aids to decision-making when the objectives are quantifiable. However, the degrees of approximation are considerable, and notably those linked to risks.

These different methods provide a snapshot of the situation: their indicators are therefore static. Some emerging methods are seeking to take account of temporal evolutions. This is the case, for example, of viability theory (Aubin, 1991) which questions the resilience of a system relative to sustainability constraints: 'given a set of constraints and a set of initial situations, is there a policy that enables compliance with the same sustainability constraints even though the data evolve?' Under this method, the series of initial, viable situations are qualified as a viability kernel. From knowledge of this kernel it is possible to deduce regulation mechanisms (and indicators); modelling then aims to clarify decision choices.

10.4.3   *Retrospective assessments of food uses and resources*

FAO data published since the middle of the twentieth century have enabled assessments, by product and by country, of the levels and trends of food consumption patterns throughout the world. The production plus imports and minus exports of a country enable an evaluation of national agricultural resources which, after the extraction of non-food uses (seeds, animal feed, biofuels, etc.) and division by the number of inhabitants, leads to an assessment of average 'food availability' per person. This availability is higher than the quantities actually ingested because it includes what has been lost after being made available to the consumer. The calorie level and content of national resources can clarify the pressure on these resources regarding the demand for food products, and notably the land necessary (within or outside the country) to meet this demand.

The *Agrimonde* foresight study (Paillard *et al.*, 2010) used these FAO data (Commodity Balances, or *CDU: Comptes Disponibilités Utilisation*) and other data sources to assess the past (1961–2003) and simulate the future (2050) of the balances between food biomass resources and their uses expressed in calories, with products grouped into five compartments according to their origin (plants, ruminants, non-ruminants, freshwater and marine fish (cf. 'Agribiom' model for greater detail; Paillard *et al.*, 2010).

The choice of the kilocalorie as the unit for food assessments facilitates the reasoning and representation of trends at highly aggregated levels (macro-regions, world) because by using calories, it is possible to sum any types of food product (plant, animal products from ruminants or non-ruminants, freshwater and marine fish). This model also enables to take account of different sources of livestock feed (food products and grasslands, in particular) to highlight the importance of animal products in the use of plant resources and, of course, of cultivable land. On the other hand, this quantification does not simulate supply and demand behaviours via prices, and does not couple production/consumption modes with their consumption of fossil fuels, water, pesticides and other elements. In *Agrimonde* (Paillard *et al.*, 2010), the assessment of the environmental and social impacts of different diets was based on expert opinions and data from the literature.

10.4.4 *Physical and synthetic indicators of environmental pressures*

Other methods have focused on converting environmental impacts into a synthetic or composite indicator. The nature of this indicator varies. The conversion of impacts into 'carbon' (CO_2) costs is currently attracting much attention, following the IPPC work on climate change.

Carbon footprint as defined by PASS 2050[1] or by the *Bilan Carbone®* (carbon balance, www.ademe.fr/bilan-carbone) is a synthetic indicator of greenhouse gas emissions. Indirectly, it also assesses the energy consumed by human activities. Frequently employed, it can be adapted to different scales, but remains limited to greenhouse gas emissions. This balance, or carbon footprint of human pressures on the climate, was selected as one of the 15 sustainable development indicators associated with the French National Strategy on Sustainable Development (SNDD) 2010–13 (Commissariat Général au Développement Durable, 2010). It has also been taken up by major retail groups and is used for the environmental labelling of consumer products (Casino, Leclerc, Tesco).

Emergy, also referred to as 'solar emergy' or 'embodied energy' is a methodology based on the principles of thermodynamics, and was developed by H.T. Odum (1996) in the early 1980s to take account of the energy required to obtain a product. Odum (1996) defined the concept of emergy as the solar energy available and used directly and indirectly to achieve a product or service. It is expressed in Solar Emergy Joules (or sej). It is useful to better identify energy fluxes arising from the environment, the technical sphere and the economy. Nevertheless, this method is not yet standardised and the source of certain reference values remains somewhat unclear.

Virtual water associates consumer goods with the quantity of water necessary to produce them. It is one of the rare indicators focused on water resources: the

---

[1] www.shopbsigroup.com/en/forma/PASS/PAS2050.

World Water Council used it to raise awareness as to the difference in water consumption between a meat-rich diet (5400 l of virtual water per day) and a vegetarian diet (2600 l) among American food consumers (Hoekstra, 2002).

The *ecological footprint* (www.footprintnetwork.com) converts the amount in tonnes of products necessary for the life of a population into hectares ('biocapacity'). The basic principle of the calculation is based on the use of a mean global yield applied to national production according to its physical (sunshine, water and land) and human characteristics. The ecological footprint – or the surface area obtained – enables a comparison of data that are not otherwise comparable. This synthetic indicator, used by the scientific community, has acquired political scope and a certain popularity. Its pedagogical value is unquestionable at a global scale (van den Bergh and Verbruggen, 1999), but it comes in for criticism (Ayres, 2000; Levett, 1998; van den Bergh and Verbruggen, 1999; among others) as soon as smaller scales are considered. Indeed, the ecological footprint does not qualify impacts. Nor is it always transparent and complete: for example, the extraction of mineral resources from the subsoil, or the extraction of fresh water, are excluded from the calculations (Boutaud and Gondran, 2009).

Some methods estimate the environmental cost of developing activities or production. The *Environmentally Weighted Material Consumption* method (Van der Voet *et al.*, 2005) developed by the European Commission proposes an indicator that combines the environmental impacts of natural resource use with European economic growth. This method groups LCAs with the domestic consumption of the EU by categories of product. It produces scores, but without being able to qualify the levels that might be sustainable.

The World Bank proposes a monetary index of sustainability called 'adjusted net saving', which integrates the physical assets, human assets and natural resources traded on markets.

These synthetic indicators are criticised when they constitute a single indicator, because whichever indicator is envisaged, the grouping of disparate data is not necessarily successful and cannot reflect both the complexity of human activities, quality of life and its sustainability. These indicators also have the drawback of being normative, because they group together information of a heterogeneous nature and give it a score and weighting. Finally, these different indicators are static; in other words, they do not take account of the adaptation and resilience of the systems studied.

10.4.5   *Nutritional and epidemiological approaches*

Social costs in terms of public health are a major concern of modern societies and the role of diet in health attracts particular attention (under-nutrition

---

Box 10.2   Measuring the biodiversity associated with food styles

It is accepted that biodiversity is an important factor in food security and improving nutrition (Toledo and Burlingame, 2006). Three levels should be considered: ecosystems, species and intra-species genetic diversity. Biodiversity as such is still poorly understood. The impact of human activities on the functional structure of biodiversity has become a global concern, but few methods can document this. A few studies have tried to introduce biodiversity in the conceptual framework of LCA (Curran *et al.*, 2011). Jeanneret *et al.* (2008) proposed an assessment method for arable crops, grasslands and semi-natural areas in the LCA framework, but it is only applicable in the Swiss context. Studies have been initiated on the use of ecosystem areas for human activities, but are as yet little applied (Koellner and Scholz, 2008). Linking food and biodiversity appears to be a major research challenge.

---

and malnutrition, obesity, cardiovascular diseases, etc.). Although the links between diet and health are clear, their global assessment still needs to be undertaken. Epidemiology explores combinations between certain food groups and health indicators. Statistical methods then measure the correlations and identify types of 'high risk' consumers (Kesse, 2010).

Methods can be divided into those which compare consumption data with a 'health' reference, and those based on the observation of consumption behaviours. The former construct indices on the quality and variety of the diet, or scores of compliance with nutritional guidelines (e.g. Mediterranean-style diet). Construction of these scores is based on scientific knowledge or hypotheses in the field of nutrition. A recent review of the literature revealed the existence of some 20 scores, many of which are derived from four main scores: the *Healthy Eating Index* (Kennedy, 1995), the *Diet Quality Index* (Patterson *et al.*, 1994), the *Healthy Diet Indicator* (Huijbregts *et al.*, 1997) and the *Mediterranean Diet* score (Trichopoulou *et al.*, 1995). Intakes of lipids (total, saturated/mono-unsaturated, cholesterol) and the consumption of fruit and vegetables are recurring parameters. However, the disparities in data collection and the construction of scores renders a comparison between these studies difficult (fish separate from other animal products or not, weighting given to different components, account taken of the energy intake or not, etc.).

The use of epidemiological methods also involves a certain degree of subjectivity and bias that is widely discussed in the literature (e.g. choice of the number of factors to be conserved).

10.4.6   *Methods to assess obstacles to and motivations for sustainable consumption*

Very few studies have focused on the motivations and attitudes of consumers towards sustainable food systems. The role of committed consumers is indeed considered in different ways: an initial approach (Cohen, 2003) highlighted their active role in the spread of interest in the environment among other consumers. It is criticised by proponents of a more political approach, which includes the development of food labelling (Boström and Klintman, 2008). A third approach calls into question the concept of 'active consumers' and emphasises the ambivalence and routine of dietary practices. The Norwegian Institute for Consumer Research (SIFO) is a supporter of this approach. Alternative food practices would then represent a social contestation reflecting alternative values (Terragni *et al.*, 2009). The French AMAP (Associations for the Maintenance of Peasant Farming), or community-shared agriculture (CSA) in Norway may indicate a change to ecological and social consciousness. Bente Halkier (2009) noted that the consumption of environmentally friendly foods results from ambivalences, tensions and dilemmas between routine social practices and a 'conscious' commitment to the environment. This viewpoint raises questions regarding institutional initiatives that could make the consumption of sustainable foodstuffs more 'normal' and routine.

Along the same lines, the ISOE (Institute for Social Ecological Research) in Germany has analysed the way in which food styles are linked to lifestyles, and particularly the points at which changes in attitude in favour of the environment are introduced (Schultz and Stieß, 2008). Empirical data have emphasised that the motivations behind the routine integration of environmental issues in these practices are ambivalent. Numerous studies on short distribution chains (or 'local food') by different disciplines (rural sociology, geographical economics, ecology, consumerism, anthropology) have also tended to highlight the trade-off between cultural and environmental approaches: local food in reaction to the globalised organisation of trade has then led to the emergence of the neologism *glocalisation* (Holt and Amilien, 2007).

10.4.7   *Composite social and ethical indicators*

Some initiatives of international scope have focused on Corporate Social Responsibility (CSR) and developed appropriate reference indicators. We should mention here the *Global Reporting Initiative* (GRI, www.globalreporting.org)

reference, and the International Labour Organisation reference,[2] ISO 26000 standard[3] or the SA 8000 standard[4] developed by Social Accountability International, a private American organisation that is also based on worker conventions and rights defined by the ILO (child labour, discrimination, freedom of association, working times, salaries, etc.). These tools, based on consultations with stakeholders, cover the three areas of sustainable development, including governance. They are designed to measure the impacts of organisations rather than processes and activities.

In the same spirit, criticism of GDP (Gross Domestic Product) has encouraged the international community to use other indicators to measure the well-being of populations. *The Human Development Index* (HDI) proposed in 1990 by Amartya Sen, Mahbub ul Haq and the UNDP is a good example. More recently, the *Multidimensional Poverty Index*, or MPI, was created at Oxford University in 2010 (Alkire, 2010; Alkire and Santos, 2010), and is also backed by the United Nations Development Programme (UNDP). The MPI approach is intermediate, between considering that a person is poor when deprived of a particular dimension, and accounting for all dimensions of deprivation (income, health, education, etc.). Published in 104 developing countries, it uses 10 indicators to assess the poverty threshold, the incidence of poverty (percentage of 'multidimensionally poor' people), and the degree of poverty (mean proportion of weighted deprivations from which a poor person suffers).

These approaches, based on quantitative and qualitative criteria and leading to this type of multidimensional index, could usefully inspire methods to assess the sustainability of food systems.

10.4.8   *Increasingly recommended multicriteria approaches*

This panorama demonstrates the growing importance of directly addressing different aspects of sustainability by means of analyses that use indicators concerning different subjects, with different scales and which can be quantitative or qualitative. It is therefore crucial to find a balance between covering the fields of sustainable development more broadly using numerous indicators, and restricting the scope of information in order to retain an intelligible vision of the situation … and then be able to issue advice or make a decision. This is the challenge of multicriteria approaches.

Several data management steps are generally employed: standardisation makes it possible to establish a relationship between different indicators and

---

[2] www.ilo.org/global/lang–en/index.htm, 2009 report on ILO indicators: www.ilo.org/. wcmsp5/groups/public/ed_emp/documents/publication/wcms_110512.pdf.
[3] www.iso.org/wgsr.   [4] www.Sa-intl.org/.

a common reference point so that they can be considered from a common standpoint; weighting allows importance to be given to indicators according to their relevance; finally, aggregation assembles the indicators around a common notion so as to limit the number of notions included in the assessment. For example, this is the case for the nesting of Principles, Criteria and Indicators: *Principle* (which concerns concepts or values), *Criteria* (which break down the Principles into operational units) and *Indicators* (which measure the criteria). The two levels of aggregation combine the information of the lower level. Depending on the types of indicators and methods used, information will be expressed numerically or graphically: green, amber or red lights, for example.

It is possible to select and combine certain indicators in order to develop a composite indicator that will then try to take account of the multidimensional nature of the subject under study. On the basis of the example of the Multidimensional Poverty Index, it may be possible to propose an analytical framework for diet sustainability (Table 10.3) linked to the different dimensions

Table 10.3 *Proposed analytical framework to construct a multidimensional indicator for the sustainability of food systems.*

Dimensions	Field	Indicators
Health, nutrition	Under-nutrition	Calories available; macro-nutritional balance; infantile under-nutrition, etc.
	Overweight, obesity	Body Mass Index, etc.
Economics	Accessibility	Food/total expenditure; Prices (values, variability, etc.); market development, etc.
	Jobs	Hours worked, etc.
Hedonism	Animal products	Animal/total proteins, etc.
	Variety of choice	Variety index, etc.
Cultural	Local production	Imports/total consumption, etc.
Religious	Respect for taboos	
Environment	Carbon	Carbon footprint, etc.
	Water	Water footprint; nitrate contents, etc.
	Biodiversity	Soil fertility; Potassium; Phosphates, etc.
Ethics	Living conditions	Human Development Index, etc.
	Equality/Fairness	GINI index; value chain, etc.

(challenges) identified previously (cf. Table 10.1). Nevertheless, the construction of such a framework is less simple than it may appear, because this approach requires consistency between the indicators using this approach. The challenge first of all is therefore to define the subject of the study in such a way that indicators from fields as diverse as health, the environment or ethics are applicable within a coherent scientific framework.

Even if they are of a multicriteria type, methods to analyse sustainability are not easy to use by decision-makers: deciding between indicators with respect to contradictory responses, or arbitrating between notions concerning different fields requires a degree of judgement that is not the job of assessment methods per se.

10.4.9   *Involving stakeholders in the assessment of sustainability also promotes changes to practices*

As we have seen above, it is important to rank indicators in the light of social choices. From the standpoint of participation, this implies debate over these choices, and also tools for their assessment. This new approach for research does not mean that scientists will stand back and leave the field open for politicians, but rather it should encourage scientists to take account, in their approach, of the potential users of their work.

Although in the past they were only a subject of study in assessments, the actors, participants and, in a broader sense, stakeholders are now regularly involved in assessment systems. In most cases, their opinions are sought regarding the objectives of the assessment, the interpretation of results and the formulation of opportunities for action.

Their intervention does not resolve issues relative to methodological limits; nevertheless, it offers an additional framework (opinions) by ranking, weighting or contextualising scientific assessments.

However, the participation of stakeholders is unanimously accepted: the representativeness and legitimacy of those convened, and the status of the positions they adopt are recurrent questions asked of participative methods: do stakeholders express their personal views or the views of the institution or more or less formal group they represent? Similarly, poor group management or only a few people monopolising a debate can diminish the relevance of the results obtained.

Despite these remarks, involving stakeholders in the definition of assessment methods can have knock-on effects that go beyond the assessment. For example, the EVAD project (Rey-Valette et al., 2008) showed that 'an approach involving the joint construction of indicators promotes organisational learning and facilitates agreement', and for this reason, it is easier to construct sustainable development practices that are shared by different levels of stakeholders if they have been analysed using concerted approaches (Rey-Valette et al., 2008). Studies in Management Sciences (Pesqueux, 2006) have generally demonstrated that assessment systems are important drivers of organisational learning. The assessment is then not only envisaged as an objective, but also as a means of driving change in organisations, territorial communities and/or among individuals.

These participative approaches have rarely been used at the scale of a complex food chain (nor indeed on diets). The enormous number of people involved becomes a major drawback and methods for the consultation and intervention of different stakeholders need to be developed. In this respect, the contribution of the social sciences to stabilising methodologies that will be both robust and operational seems crucial. Note should however be made of the existence of some tools that may serve the selection of indicators or the combination of expert opinions, arising notably from Operational Research. Particular mention should be made of the ELECTRE method (*ELimination Et Choix TRraduisant la rEalité*, or elimination and choice reflecting reality) which can sort, rank and choose indicators using a weighting approach (Roy, 1968); the Delphi method which enables a convergence of expert opinions (Rowe and Wright, 1999), or the use of multi-attribute models to break down decision-making, such as the DEXi method (Bohanec et al., 2000).

10.5   Questions for research

It is practically impossible to list all the research questions associated with analytical methods and the sustainability of food styles, in view of the number of methods and the multiplicity of subjects and challenges. We list below some of the questions we consider to be important, particularly concerning LCA and associated methods (because of their broad application).

10.5.1   *The choice of an indicator system*

An indicator system is the toolbox that enables the assessment and then the orientation of the system under study. The set of indicators must be consistent; they must complement each other and be independent. They must be sufficiently numerous to cover all the fields deemed strategic to the assessment, and be few enough to be interpretable. Depending on the objective of the study and its geographical scope of application, the choice of indicators may differ. But who determines the choice of these indicators? Is the same type of response obtained by using a normative and proven indicator system or by enabling the joint construction of an indicator system with representatives from different levels of stakeholders? In the latter case, how should the levels of stakeholders to be involved in such a complex system as a food system, with overlapping spatial scales, be chosen?

10.5.2   Complementary indicators

Some of the pillars of sustainable development are deficient in terms of operational indicators.

The arsenal of existing indicators is not yet sufficient to answer all the questions posed. Particular reference can be made to the case of biodiversity, which is the subject of regular demands from civil society. How can the notion of biodiversity be integrated in the analysis of food systems? What levels of biodiversity should be taken into account (genotypes, species, ecosystems)? Is it possible to aggregate them? How can they be aggregated at different geographical levels?

Regarding existing indicators, conceptual adaptations need to be made to approach the sustainability of food systems. This is the case for the different approaches to the consumption or use of water.

10.5.3   The functional unit

LCA is applied to a function or product (hence the term, functional unit). The calculations for all indicators refer to this unit. What is the functional unit that best reflects the quality of a nutrition style (calories, proteins, omega 3, etc.)? Can functional units be proposed that combine different aspects? What is the influence of the choice of functional unit on the results and the significance of analyses at the scale of food systems?

10.5.4   Recycling and allocations in food systems

Different studies at the scale of food systems have demonstrated losses throughout the chain. It is necessary to determine the fate of these losses and to understand how they can contribute to improving the economic, social and environmental balance of food systems, by means of recycling or re-use phenomena external to the system (particularly in developing countries). The choice of rules on the allocation of impacts between the main products and different by-products therefore becomes crucial, as is also the case in production systems with various end-products (e.g. between milk, cull cows and veal; see Cederberg and Stadig, 2003). Should we follow mass-based rules (pro rata of physical properties) on certain characteristics (e.g. energy content) or on economic rules (value of different substitution products)? How do these choices influence the results (e.g. partial or total allocation of an impact avoided by a substitution process)? How can a uniform approach be designed that combines the functional unit and allocation rules, to cover the entire food system?

10.5.5   The case of social LCA

Although economic evaluations can be rendered compatible with LCA because they are based on monetary flows (cf. Life Cycle Costing), the issue of the integration of social aspects in LCA remains open. Several levels of questions are posed: what precisely are the epistemic foundations for the social assessment of a life cycle? What is the consensus concerning inventory indicators? How can social indicators be related to a functional unit?

10.5.6   Scales and representativeness of indicators

How can the local context be included in assessing the relevance of certain indicators (definition of contextualised indicators, integration of the sensitivity of recipient environments when calculating impacts and the availability of resources)? This is necessary regarding the consumption of resources (e.g. water) where levels of availability are essential in determining their value (geographical and temporal scales) or for indicators on emissions (e.g. eutrophication) where the sensitivity of the recipient environment is important to determine impact. Parallels are possible in the economic sphere. Combinations of methods such as LCA and risk analysis are possible, but have not yet been developed.

10.5.7   Uncertainty

The input data used to determine indicator values are obtained from various sources and are often heterogeneous. They are often based on surveys or published data, statistics, field measurements or descriptions of specific cases (individuals, systems, etc.). To all these data are attached a level of uncertainty linked to the observation and measurement methods used. These uncertainties propagate throughout the system under study and will influence the quality of the results. The degree of uncertainty is often neglected in multicriteria analyses, and the consequences for decision-making are probably underestimated. In complex systems such as food styles or food systems, the methods used to calculate and analyse this propagation are still difficult to implement and need to be studied in greater depth.

10.5.8   Data availability and infrastructure requirements

The studies referred to in this document are based on a mass of information on different scales: world trade, countries or regions, sectors, farms, individuals (such as consumers), etc.

Information on many aspects studied is not available at these different scales, and the absence of reliable, verified and documented data is a major obstacle to the conduct of scientific work. For this reason, there is a real need for shared observations (on consumption or behaviour), statistics and databases on products, sectors, and processing and marketing methods (such as those which can be used for LCA). This is the case in developed countries, and it is crucial in developing countries, where the need for observation is even greater because the initial situation is even less documented. In many studies, recourse to a field survey is unavoidable and must remain standard practice so as to ensure the quality of data and to accurately understand the different contexts (physical, social, economic, etc.). As in any other scientific discipline, an assessment of sustainability requires the collection and evaluation of data quality. Any research strategy in this area which is solely based on the processing of pre-existing data is highly questionable.

10.6   Conclusion

These discussions on the methods required to analyse the sustainability of food systems were rapidly confronted with the immensity of the field. Food is a multidimensional object; although efforts have been made to simplify it by means of market logics, it is necessary to understand it as a complex social phenomenon (Dhérissard and Viel, 2007).

For this reason, an assessment through the challenges seems most appropriate because this approach allows a choice of assessment tools according to trade-offs by society, the sensitivity of a region to environmental pressures, the perspectives for global change, etc. Indeed, the actual definition of these challenges poses a question: who can legitimately define them? And how can scientists or assessors position themselves with respect to the concerns of society?

The approach preferred here consisted in considering assessment methods based on a grid to structure the challenges of food sustainability. This approach differed, first, from those based on the opinions of experts which address food-related questions without taking account of assessment methods, as seen for example in the work by the FAO (Pretty *et al.*, 2010). Second, it also differed from studies focused on methods, such as the CALCAS synthesis on LCAs (Zamagni *et al.*, 2009) and its derivative methods, which do not deal with the issue of food. For this reason, this chapter, although summary, offers an original view on the issue under study.

The trend in assessment methods is thus towards a 'multi' assessment: multi-criteria, multidimensional, multi-approach, etc. Indeed, no assessment of sustainability can be limited to one standpoint, and an amalgam must be made between methods: static or dynamic, mathematical or participative. The scales to be considered, whether temporal or geographical, are also multiple, as are the stakeholders who need to be involved. The need thus emerges for cross-disciplinary studies, the frameworks for which are as yet unknown, including on the links that need to be built between research and the worlds of production, industry, distribution, consumption and development. These partnerships will both determine the provision of relevant, reliable data on food systems, which is crucially lacking at present, and determine the value systems (ethics) in the context of which any assessments of sustainability must be made.

Economic or environmental rationales alone are therefore not sufficient to respond to such complex challenges, and it is possible that research will advance towards disruptions in thinking. In this vein, Elghali et al. (2008) stated that 'the loss of certainty and the intrusion of ethics destroy the basis for a "normal" scientific approach'.

References related to Table 10.2

References with an indication of the method used (cf. Table 10.2: examples of methods and indicators that respond to the challenges or issues of sustainability).

Human development indicator (HDI)

Nourry M. 2008. Measuring sustainable development: some empirical evidence for France from eight alternative indicators. *Ecological Economics*, 67 (3): 441–456. Available at: http://dx.doi.org/10.1016/j.ecolecon.2007.12.019.

Rosenbloom, J. I., Kaluski, D. N. and Berry, E. M. 2008. A global nutritional index. *Food and Nutrition Bulletin*, 29 (4): 266–277. Available at: http://www.ncbi.nlm.nih.gov/pubmed/19227051.

Shah, Z. and Kumar, M. D. 2008. In the midst of the large dam controversy: objectives, criteria for assessing large water storages in the developing world. *Water Resources Management*, 22 (12): 1799–1824. Available at: https://collaboratif.inra.fr/silverpeas/attached_file/componentId/kmelia20980/documentId/73255/versionId/95536/name/10.1007/s11269-008-9254-8.

Mass balance

Ayres, R. U. and Ayres, L. W. 1998. *Accounting for Resources, 1: Economy-wide Applications of Mass-balance Principles to Materials and Waste*. Cheltenham, UK: Edward Elgar Publishing, 245 pp.

Barbiero, G., Camponeschi, S., Femia, A. *et al.* 2003. 1980–1998 material-input-based indicators time series and 1997 material balance of the Italian economy. Rome: ISTAT (Institut Nazinale di Statistica).

Giljum, S., Hubacek, K. and Sun, L. 2004. Beyond the simple material balance: a reply to Sangwon Suh's note on physical input–output analysis. *Ecological Economics*, 48 (1): 19–22. Available at: http://dx.doi.org/10.1016/j. ecolecon.2003.09.004.

Materials balance

Barbiero, G., Camponeschi, S., Femia, A. *et al.* 2003. 1980–1998 material-input-based indicators time series and 1997 material balance of the Italian economy. Rome: ISTAT (Instituto Nazinale di Statistica).

Giljum, S., Hubacek, K. and Sun, L. 2004. Beyond the simple material balance: a reply to Sangwon Suh's note on physical input–output analysis. *Ecological Economics*, 48 (1): 19–22.

United Nations 1976. Draft guidelines for statistics on materials/energy balances, document E/CN.3/493. New York: United Nations.

Input/output

Beutel, J. 2008. *Eurostat Manual of Supply, Use and Input-Output Tables. Eurostat Methodologies and Working Papers.* Luxembourg: Office for Official Publications of the European Communities, 592 pp. Available at: http://epp.eurostat.ec.europa. eu/cache/ITY_OFFPUB/KS-RA-07-013/EN/KS-RA-07-013-EN.PDF.

Bullard, C. W., Penner, P. S. and Pilati, D. A. 1978. Net energy analysis: handbook for combining process and input–output analysis. *Resources and Energy*, 1 (3): 267–313. Available at: http://dx.doi.org/10.1016/0165-0572(78)90008-7.

Duchin, F. 1992. Industrial input–output analysis: implications for industrial ecology. *Proceedings of the National Academy Science USA*, 89: 851–855. Available at: http://dx. doi.org/10.1073/pnas.89.3.851.

European Commission Eurostat home page, ESA 95: Supply, use and input–output tables. Available at: http://epp.eurostat.ec.europa.eu/portal/page/portal/ esa95_supply_use_input_tables/introduction.

European Commission 2001. European Governance. A White Paper. COM(2001) 428 final. Brussels.

European Network of Environmental Input–Output Analysis 2001. Proceedings of 1st Meeting, 8 May 2001, Madrid, Spain.

Finnveden, G. and Moberg, Å. 2005. Environmental systems analysis tools: an overview. *Journal of Cleaner Production*, 13: 1165–1173. Available at: http://dx.doi. org/10.1016/j.jclepro.2004.06.004.

Grêt-Regamey, A. and Kytzia, S. 2007. Integrating the valuation of ecosystem services into the Input-Output economics of an Alpine region. *Ecological Economics*, 63: 786–798. Available at: http://dx.doi.org/10.1016/j.ecolecon.2007.02.026.

Keuning, S. J. and de Ruijter, W. A. 1988. Guidelines to the construction of a social accounting matrix. *Review of Income and Wealth*, 34 (1): 71–100. Available at: https://collaboratif.inra.fr/silverpeas/attached_file/componentId/ kmelia20980/ documentId/73255/versionId/95536/name/10.1111/j.1475-4991.1988.tb00561.x.

Machado, G., Schaeffer, R. and Worrell, E. 2001. Energy and carbon embodied in the international trade of Brazil: an input–output approach. *Ecological Economics*, 39 (3): 409–424. Available at: http://dx.doi.org/10.1016/S0921-8009(01)00230-0.

Munksgaard, J., Wier, M., Lenzen, M. and Dey, C. 2005. Using input–output analysis to measure the environmental pressure of consumption at different spatial levels. *Journal of Industrial Ecology*, 9 (1–2): 169–185. Available at: http://dx.doi.org/10.1162/1088198054084699.

Pietroforte, R., Bon, R. and Gregori, T. 2000. Regional development and construction in Italy: an input–output analysis, 1959–1992. *Construction Management and Economics*, 18 (2): 151–159. Available at: http://dx.doi.org/10.1080/014461900370780.

Suh, S. 2005. Developing a sectoral environmental database for input-output analysis: the Comparative Environmental Data Archive of the US. *Economic Systems Research*, 17 (4): 449–469.

Weisz, H. and Duchin, F. 2006. Physical and monetary input–output analysis: what makes the difference? *Ecological Economics*, 57 (3): 534–541. Available at: http://dx.doi.org/10.1016/j.ecolecon.2005.05.011.

Life cycle assessment (LCA)

Guinée, J. B. E. 2002. *Handbook on Life Cycle Assessment: Operational Guide to the ISO Standards*. Dordrecht, The Netherlands: Kluwer Academic Publishers, 692 pp.

Hubacek, K. and Giljum, S. 2003. Applying physical input–output analysis to estimate land appropriation (ecological footprint) of international trade activities. *Ecological Economics*, 44 (1): 137–151. Available at: http://dx.doi.org/10.1016/S0921-8009(02)00257-4.

Wackernagel, M. and Rees, W. 1996. *Our Ecological Footprint. Reducing Human Impact on the Earth*. Gabriola Island, Canada: New Society Publishers, 160 pp.

Water footprint

Anon. 2010. Australia's food industry reduces carbon and water footprint. *Food Australia*, 62 (4): 153.

Bulsink, F., Hoekstra, A. Y. and Booij, M. J. 2010. The water footprint of Indonesian provinces related to the consumption of crop products. *Hydrology and Earth System Sciences*, 14 (1): 119–128. Available at: http://doc.utwente.nl/76918/1/Bulsink10water.pdf.

Canals, L. M. I., Chapagain, A., Orr, S. *et al.* 2010. Assessing freshwater use impacts in LCA, part 2: case study of broccoli production in the UK and Spain. *International Journal of Life Cycle Assessment*, 15 (6): 598–607. Available at: http://dx.doi.org/10.1007/s11367-010-0187-0.

Ridoutt, B. G. and Pfister, S. 2010. A revised approach to water footprinting to make transparent the impacts of consumption and production on global freshwater scarcity. *Global Environmental Change: Human and Policy Dimensions*, 20 (1): 113–120. Available at: http://dx.doi.org/10.1016/j.gloenvcha.2009.08.003.

Zeitoun, M., Allan, J. A. and Mohieldeen, Y. 2010. Virtual water 'flows' of the Nile Basin, 1998–2004: a first approximation and implications for water security. *Global Environmental Change: Human and Policy Dimensions*, 20 (2): 229–242. Available at: http://dx.doi.org/10.1016/j.gloenvcha.2009.11.003.

Carbon footprint

BSI (2008) PAS 2050 Specification for the assessment of the life cycle greenhouse gas emissions of goods and services: under consultation.

Carbon Trust 2006. Carbon footprints in the supply chain: the next step for business. Report Number CTC616, The Carbon Trust, London, UK, 24 pp. Available at: http://www.iee-library.eu/images/all_ieelibrary_docs/21 carbon footprints in the supply chain the next step for business.pdf.

Carbon Trust 2007a. Carbon footprinting: an introduction for organisations, 6 pp. Available at: http://www.carbondecisions.ie/resources/ footprint_for_organisations.pdf.

Carbon Trust 2007b. Carbon Trust Labelling Scheme rolls out across the country. Available at: http://www.carbontrust.co.uk/news/news/press-centre/2007/Pages/ 260707_CT_label.aspx.

European Commission 2007. Carbon footprint: what is it and how to measure it. Available at: http://lct.jrc.ec.europa.eu/pdf-directory/Carbon-footprint.pdf.

IEC 2008. Introduction, intended uses and key programme elements. The International EPD Cooperation. Available at: http://www.environdec.com/ Documents/GPI/EPD_introduction_080229.pdf.

ISO 2006. Environmental labels and declarations. Type III environmental declarations: principles and procedures. IS14025:2006.

Johnson, E. 2008. Disagreement over carbon footprints: a comparison of electric and LPG forklifts. *Energy Policy*, 36 (4): 1569–1573. Available at: http://dx.doi.org/ 10.1016/j.enpol.2008.01.014.

Wiedmann, T. and Minx, J. 2007. A Definition of Carbon Footprint, ISA UK Research Report 07–01, 11 pp. Available at: http://www.censa.org.uk/docs/ISA-UK_Report_07-01_carbon_footprint.pdf.

Yu, Y., Hubacek, K., Feng, K. and Guan, D. 2010. Assessing regional and global water footprints for the UK. *Ecological Economics*, 69 (5): 1140–1147. Available at: http:// dx.doi.org/10.1016/j.ecolecon.2009.12.008.

Porter value chain

Teweldemedhin, M. Y. 2008. The fish industry in Eritrea: from comparative to competitive advantage. *African Journal of Agricultural Research*, 3 (5): 327–333.

Koivisto, A. and Varnik, R. 2005. Competitiveness of strawberry production in Finland and Estonia. Rural Development 2005, Vol. 2, Book 1, Proceedings – Globalisation and Integration Challenges to Rural Development in Eastern and Central Europe. Kauno Raj, Lithuanian Univ Agriculture, pp.127–129. Available at: http://www.lzuu.lt/rural_development/archive/2005/CD/Papers in PDF/S2_PDF/ S2_12.pdf.

Life cycle costing (LCC)

Ala-Risku, T. and Kopri, E. 2008. Life cycle costing: a review of published case studies. *Managerial Auditing Journal*, 23 (3): 240–261. Available at: http://dx.doi.org/10.1108/02686900810857703.

ASTM E917–02 Standard practice for measuring life-cycle costs of buildings and building systems. Available at: http://www.astm.org/Standards/E917.htm.

Gluch, P. and Baumann, H. 2004. The life cycle costing (LCC) approach: a conceptual discussion of its usefulness for environmental decision-making. *Building and Environment*, 39: 571–580. Available at: http://dx.doi.org/10.1016/j.buildenv.2003.10.008.

Ciroth, A., Hunkeler, D., Huppes, G. *et al.* 2008. *Environmental Life Cycle Costing*. Pensacola, FL: SETAC Press, Taylor and Francis.

Hunkeler, D., Lichtenvort, K. and Rerbitzer, G. (eds) 2008. *Environmental Life Cycle Costing*. Pensacola, FL: SETAC-CRC Press, 232 pp.

International Electrotechnical Commission 2004. IEC 60300–3–3: Dependability Management. Part 3–3: Life cycle cost analysis. Application guide.

Ness, B., Urbel-Piirsalu, E., Anderberg, S. and Olsson, L. 2007. Categorising tools for sustainability assessment. *Ecological Economics*, 60: 498–508. Available at: http://dx.doi.org/10.1016/j.ecolecon.2006.07.023.

Rebitzer, G. and Seuring, S. 2003. Methodology and application of life cycle costing. *International Journal of Life Cycle Assessment*, 8 (2): 110–111. Available at: http://dx.doi.org/10.1007/BF02978436.

Schmidt, W-P. 2003. Life cycle costing as part of design for environment, environmental business cases. *International Journal of Life Cycle Assessment*, 8 (3): 167–174. Available at: http://dx.doi.org/10.1007/BF02978464.

Udo de Haes, H., Heijungs, R., Suh, S. and Huppes, G. 2004. Three strategies to overcome the limitations of life-cycle assessment. *Journal of Industrial Ecology*, 8 (3): 19–32. Available at: http://dx.doi.org/10.1162/1088198042442351.

Contribution to wealth

Gadrey, J. and Jany-Catrice, F. 2007. *Les nouveaux indicateurs de richesse*. Repères, n°404. Paris: La Découverte, 123 pp.

Dreyer, L., Hauschild, M. and Schierbeck, J. 2006. A Framework for Social Life Cycle Impact Assessment. *International Journal of Life Cycle Assessment*, 11 (2): 88–97. Available at: http://dx.doi.org/10.1065/lca2005.08.223.

Grießhammer, R., Benoît, C., Dreyer, L. C. *et al.* 2006. Feasibility study: integration of social aspects into LCA. Available at: http://biblio.ugent.be/input/download?func=downloadFile&fileOId=633083.

Hunkeler, D. 2006. Societal LCA methodology and case study. *International Journal of Life Cycle Assessment*, 11 (6): 371–382. Available at: http://dx.doi.org/10.1065/lca2006.08.261.

Jørgensen, A., Hauschild, M. Z., Jorgensen, M. S. and Wangel, A. 2009. Relevance and feasibility of social life cycle assessment from a company perspective. *International Journal of Life Cycle Assessment*, 14 (3): 204–214. Available at: http://dx.doi.org/10.1007/s11367-009-0073-9.

Jørgensen, A., Le Bocq, A., Nazarkina, L. and Hauschild, M. 2008. Methodologies for social life cycle assessment. *International Journal of Life Cycle Assessment*, 14 (2): 96–103. Available at: http://dx.doi.org/10.1065/lca2007.11.367.

Norris, G. A. 2006. Social impacts in product life cycles: towards life cycle attribute assessment. *International Journal of Life Cycle Assessment*, 11 (Special Issue 1): 97–104. Available at: http://dx.doi.org/10.1065/lca2006.04.017.

Weidema, B. P. 2005. ISO 14044 also applies to social LCA. *International Journal of Life Cycle Assessment*, 10 (6): 381. Available at: http://dx.doi.org/10.1065/lca2005.11.002.

Weidema, B. P. 2006. The integration of economic and social aspects in life cycle impact assessment. *International Journal of Life Cycle Assessment*, 11 (Special Issue 1): 89–96. Available at: http://ftp.jrc.es/EURdoc/JRC46650.pdf.

Global reporting initiative (GRI)

Global Reporting Initiative (GRI) Available at: http://www.globalreporting.org/.

Corporate social responsibility (CSR)

Social accountability international Available at: http://www.sa-intl.org/.

Observatoire sur la Responsabilité Sociétale des Entreprises Available at: http://www.orse.org/.

ISO Social responsibility Available at: http://www.iso.org/iso/social_responsibility.

ILO standards

International Labour Organization Available at: http://www.ilo.org/.

Cost–benefit analysis

Abt 2000. The particulate-related health benefits of reducing power plant emissions. October 2000. Prepared for EPA by Abt Associates Inc., 4800 Montgomery Lane, Bethesda, MD 20814-5341, 206 pp. Available at: http://www.abtassociates.com/reports/particulate-related.pdf.

European Commission 2006. The CAFE Programme (Clean Air for Europe). European Commission DG Environment. Available at: http://www.airclim.org/policy/sub6_10.php.

ExternE 2005. ExternE. Externalities of energy: methodology 2005 Available at: http://ec.europa.eu/research/energy/pdf/kina_en.pdf.

Holland, M., Watkiss, P., Pye, S., de Oliveira, A. and van Regemorter, D. 2005. Cost–benefit analysis of policy option scenarios for the Clean Air For Europe Programme. Available at: http://www.cafe-cba.org/assets/thematic_strategy_analysis_v3.pdf.

McKone, T. E. and Enoch, K. G. 2002. CalTOX™, a multimedia total exposure model. Report LBNL: 47399. Lawrence Berkeley National Laboratory, Berkeley, CA. Available at: http://www.escholarship.org/uc/item/9j74302f.pdf.

Mitchell, R. C. and Carson, R. T. 1989. *Using Surveys to Value Public Goods: the Contingent Valuation Method.* Washington DC: Resources for the Future, 463 pp.

Pearce, D. 2001. Integrating cost-benefit analysis into the policy process, annex II in Valuing the benefits of environmental policy. RIVM report 481505 024, Bilthoven.

Rabl, A. 2000. Criteria for limits on the emission of dust from cement kilns that burn waste as fuel. ARMINES/Ecole des Mines de Paris, Paris. March 2000, 10 pp. Available at: http://www.bvsde.paho.org/bvsacd/cd24/rabl.pdf.

Rabl, A., Spadar, J. V. and van der Zwaan, B. 2005. Uncertainty of air pollution cost estimates: to what extent does it matter? *Environmental Science & Technology*, 39 (2): 399–408. Available at: http://dx.doi.org/10.1021/es049189v.

Skovgaard, M., Ibenholt, K. and Ekvall, T. 2007. Nordic guideline for cost–benefit analysis in waste management. TemaNord 2007:574. Nordic Council of Ministers, Copenhagen. Available at: http://www.norden.org/da/publikationer/publikationer/2007-574/at_download/publicationfile.

Spadaro, J. and Rabl, A. 2008. Estimating the uncertainty of damage costs of pollution: a simple transparent method and typical results. *Environmental Impact Assessment Review*, 28 (2): 166–183. Available at: http://dx.doi.org/10.1016/j.eiar.2008.08.003.

Spadaro, J. and Rabl, A. 2004. Pathway analysis for population-total health impacts of toxic metal emissions. *Risk Analysis*, 24 (5): 1121–1141. Available at: http://web.me.com/arirabl/Site/Publications_files/Spadaro+Rabl04 Metals.pdf.

Risk assessment

Calow, P. 1998. *Handbook of Environmental Risk Assessment and Management*. Oxford: Blackwell Publishing, 82 pp.

Cowell, S. J., Fairman, R. and Loftstedt, R. E. 2002. Use of risk assessment and life cycle assessment in decision making: a common policy research agenda. *Risk Analysis*, 22 (5): 879–894. Available at: http://dx.doi.org/10.1111/1539-6924.00258.

DEFRA (UK Department of the Environment, Transport and the Regions) 2000. Guidelines for environmental risk assessment and management. Available at: http://archive.defra.gov.uk/environment/quality/risk/eramguide/.

Fairman, R., Williams, W. and Mead, C. 1998. Environmental risk assessment: approaches, experiences and information sources. *Environmental Issues Series*, No. 4, European Environment Agency, Copenhagen.

Flemströ, K., Carlson, R. and Erixon, M. 2004. Relationships between life cycle assessment and risk assessment: potentials and obstacles. Naturvårdsverket Report 5379, 82 pp. Available at: http://www.naturvardsverket.se/Documents/publikationer/620-5379-5.pdf.

Olsen, S., Christensen, F., Hauschild, M. *et al.* 2001. Life cycle impact assessment and risk assessment of chemicals: a methodological comparison. *Environmental Impact Assessment Review*, 21 (4): 385–404. Available at: http://dx.doi.org/10.1016/S0195-9255(01)00075-0.

Saouter, E. and Feijtel, T. 2000. Use of life cycle assessment and environmental risk assessment in an integrated product assessment. In M. Hauschild, S. Olsen, C. Poll and F. Bro-Rasmussen, eds. *Risk Assessment and Life Cycle Assessment*. Copenhagen: TemaNord.

Sleeswijk, A., Heijungs, R. and Erler, S. 2003. Risk assessment and life-cycle assessment fundamentally different yet reconcilable. *Greener Management International*, 41 (Spring): 77–87.

Conclusion

Authors: CATHERINE ESNOUF, MARIE RUSSEL
AND NICOLAS BRICAS

Because of the approaches adopted according to their links to and potential impacts on food sustainability, the discussions that took place in the context of duALIne were designed to compile an inventory of the principal determinants governing past trends in food systems, to identify the critical points in these systems with respect to environmental, social, health and economic issues, and finally to highlight research questions that could be addressed by programmes in the future. This conclusion presents the main findings of this analysis, a summary of the research themes determined, subjects that still need to be explored and the prospects offered.

Knowledge acquired and questions for research

Western-style food systems are not sustainable

It is clear from the scientific literature that Western-style food systems, and of course their global extension, are not sustainable in terms of their consumption of resources, impacts on ecosystems and effects on health (overweight, obesity and associated pathologies). The central question is thus to determine which trends, which transitions or even which ruptures could lead to the emergence of more sustainable systems (integrating both the supply of products and the demands of food consumers).

Consumer behaviour

According to a retrospective analysis, consumption can be predicted in the long term. In particular, the quantity of calories, and then the consumption of calories of animal

Food System Sustainability: Insights from duALIne, eds. Catherine Esnouf, Marie Russel and Nicolas Bricas. Published by Cambridge University Press. © Cambridge University Press 2013.

origin, increase in line with economic development to achieve stability. *This trend, at the scale of the planet, has negative effects on sustainability.* Furthermore, food transitions are developing more rapidly today (emerging countries) than they did in the past (OECD countries).

One might question the link between this increase and the satisfaction of physiological needs, but the analysis has shown that these needs are not a determining factor.

However, if we break down consumption and expenditure to the level of products within major categories, significant and persistent differences can be seen between countries that are otherwise relatively similar in terms of their economic development. For example, dairy products are consumed in markedly different ways within Europe; the weight of history and culture thus seems to be determinant.

In terms of action on behaviour, studies have tended to show that agricultural support policies in the past did not have negative effects when compared with nutritional guidelines. Policies on nutritional information are only effective if this information is validated and consensual over a long period. Preliminary studies also seem to show that, at least in the short term, we cannot rely upon consumers' willingness to pay to fund the probable higher cost of more sustainable food; their influence tends to take the form of a propensity to sanction products that do not take account of these options. These results deserve to be confirmed; they show that research is also necessary on economic models that integrate this additional cost elsewhere in the food system.

The major questions for research therefore concern the degrees and patterns of changes to consumption, as well as the impacts on sustainability of accelerating nutritional transitions in emerging countries and of the increases in food inequalities.

Food systems and chains

The dominant food system in industrialised countries has been based on reducing the range of agricultural raw materials produced and on the specialisation of operators at two successive stages: fractionation of these raw materials and then reformulation in order to generate a broad diversity of final food products of regular and satisfactory quality.

Changes to processes through the addition of successive criteria (food safety, organoleptic quality, nutritional quality) reduced industry's leeway and degree of freedom. Will it be possible to comply with the constraints of sustainability without having to go back on some of the constraints previously integrated? Is it possible to respond by optimising existing technologies, or must food processing methods and the organisation of food systems be fundamentally re-designed?

Agricultural production today faces growing constraints. These are illustrated, for example, by greater variability (notably in terms of price) of raw materials. In such a context, *food chains will have to develop more flexible systems that are also robust in the event of extreme situations.* Sustained high prices, the increasing scarcity of energy and water resources, and an accumulation of environmental constraints throughout the sector, will lead to changes to the design of processes (fractionation, phase changes, etc.) and to logistical organisation (storage, transport and the cold chain).

In this respect, a scientific foresight study on the determinants, characteristics and numerous consequences of variability in raw materials (quantities, quality and prices) needs to be carried out in the future.

Alternative systems exist, or are emerging. The question is to determine how their observed, possible or desirable growth, and their interactions with the agri-industrial system, will (or will not) contribute to greater sustainability. At present, analyses diverge regarding their contribution to sustainability. In this context, research needs to better characterise the major options possible, and, based on an integrated vision, highlight new potential patterns for different chains, according to the hypotheses and contexts of the levels of development achieved to date in different countries.

Taking account of food, energy and chemical needs requires a systemic vision of food systems. The challenge of this systemic approach is to define how it is possible to study a complex system made up of production and processing operations, with human input throughout the chain. The main question is to determine how to obtain the small quantity of data that will be crucial. In this framework, constructing experimental strategies based on an understanding of these operations appears essential and decisive. The corollary is the need to assemble and integrate knowledge so as to facilitate operational choices that reconcile the three challenges of food, energy and chemistry in a sustainable biosphere.

In particular, the development of new generations of biorefining systems that will optimise both the use of all raw materials and the production factors, for food and non-food markets, merits further exploration. The conditions for their installation in a given area must be simulated in order to minimise any damage to ecosystems, or even to contribute positively to the quality of the environment.

Any choices relative to technological and agronomic pathways must be informed so as to highlight opportunities to increase the efficiency of different strategies in light of economic, environmental (water, GHG emissions, biodiversity, limitation of losses, etc.), energy and social (employment) criteria.

In terms of methods, the links between industrial choices and productivity justify crossover investigations at the economic and technological levels.

More generally, the theoretical bases and tools required to integrate bio-logical/physical/chemical/economic processes in indicators and models at different scales (temporal, spatial or organisational), and the laws of scale change, constitute a crucial challenge.

A better understanding of the challenges linked to losses and wastage

An analysis of current research on losses and wastage revealed the amount of work that still needs to be done in order to better understand this subject and interpret its consequences in terms of sustainability. Chapter 7 highlights some potential research areas, which differ according to the region of the world and the scale considered. They can be divided into *two important areas: the reduction in losses and wastage and the use of losses and wastage in priority for human food* rather than for other uses (animal feed, energy, chemistry, recovery, etc.).

Reciprocal influences of food systems and behaviours: the case of migration

Migratory flows from Southern to Northern countries, but particularly between Southern countries, generate innovations in food systems: new products circulate, and new marketing and distribution chains develop in parallel with the penetration of multinational processing and distribution firms. This is now being observed in a very large number of countries. How, in this context, will these global and specific food chains co-exist or clash? What will be the effects on different food cultures, from uniformity to the affirmation of specific identities or innovations? Will these trends promote the nutritional status of the individuals concerned or not?

Urbanisation and location

A major challenge is the need to feed large cities in a manner that guarantees or even improves the dietary situation of rural populations. *The main point is that a changing urban society challenges the sustainability of food systems in terms of transport and supply modes, as well as modifying dietary practices.* This calls into question the possibility of local sustainable development, the introduction of urban agriculture and the travel of consumers to retail outlets. Finally, it raises questions regarding the spatial and nutritional exclusion of some populations and the resilience of supplies for urban communities in a context of major events or crises.

The questions for research raised concern an assessment of the sustainability of the different scenarios for a reconfiguration of urban food systems. They

highlight in particular the impact of energy prices on the relocation and organisation of actors in food chains and different regions in a context of growing
urbanisation. Research should thus aim to establish a link between food systems
and other systems: energy, transport and its infrastructures, urban policies and
regional development, and the spatial organisation of activities – work, residence, supplies, leisure. These efforts should therefore involve research communities from a wide variety of disciplines.

Governance

Knowledge is necessary on the characteristics of food system governance modes that will provide possible options between the two extremes of
consumer 'over-responsibility' and systematic recourse to the regulation of
supply. *The key is to identify and assess the public and private drivers of change
towards systems that take better account of the different dimensions of sustainability.*
How can territorial dynamics and industrial strategies be combined? What is
the impact of interventions by new actors, such as local communities, on the
emergence of local food systems? And finally, to what extent does the
development of more sustainable options call into question the distribution
of value as well as decision-making and prescriptive powers throughout the
value chain?

The issue of the shift between consumer information and over-responsibility,
raised in Chapters 2 and 9, combines the risks of increasing exclusion and
avoiding public debate on sustainable development. It takes on new importance
if we consider that food systems can be characterised by the flows of information that surround them – in the same way as flows of energy or matter but via
different channels: the vulnerability of systems can be considered via the vulnerability of information flows. The numerous facets of this question of information flows require joint study by experts in information sciences and
technologies and by those from the fields of economic, social and political
sciences.

International markets: a context of scarcity, the impact of standards, vulnerability and volatility

The rapid growth of demand for agricultural products for food or energy
purposes, combined with the constraints of exploiting fossil resources and
protecting the environment, may create long-term tensions in international
markets and increase the volatility of food prices. *It is therefore important to analyse
the impact of abrupt price rises on food security,* not only by considering rural
producers and urban consumers, but also by studying all food systems (and
notably the transmission of prices from international to domestic markets), *to*

assess the ability of different food systems to resist shocks and to identify the methods necessary to manage these situations.

As a corollary, it is necessary to assess the effects of the massive land acquisitions that are developing at present, and to clarify the link between this phenomenon and awareness to these long-term market tensions. Similarly, in Northern countries, the health, environmental and social demands of markets have resulted in a proliferation of quality standards. These standards can act as non-tariff barriers to trade: they also have major consequences regarding the types of agriculture adopted by developing countries, and tend to result (like the large-scale purchase of land) in the development of industrial agriculture with salaried employees. What effects will these changes have on the environment and biodiversity, on social equity and on the losses and wastage linked to non-compliance? What are the true consequences of the sustainable development targets that these standards are designed to achieve?

Impact assessment methods

The duALIne project served as a reminder that food sustainability is a multidimensional issue. It must be envisaged at different temporal and spatial scales, and must prove its robustness during scale changes: an apparently sustainable local system may in fact exert negative effects on its periphery or even on more distant food systems. *The main point is that there are numerous methods and indicators available for assessment, but taken separately they are unable to cover all the challenges of food sustainability.* Such assessment methods must have three important characteristics: they must focus on precisely defined challenges that are appropriate to the question being posed, they must take account of the vision of stakeholders and society, and they must tend towards multicriteria approaches. One method that is increasingly being used to assess environmental impact is LCA. Its methodological framework has now been extended to include economic and social dimensions. Improvements to this method, and the application of LCA to complex systems such as food, raise further research questions that are both generic regarding the implementation of assessment (functional unit, uncertainty, allocations, etc.) and applied regarding the objects of the assessment (agriculture, chains, national territorial approaches, etc.).

Multicriteria analysis can establish links between different challenges, even if the large number and different dimensions of these challenges hampering objective aggregation and of political trade-offs may prove necessary. Methods and criteria should also be developed to enable the simulation of trends affecting food systems and to assess scenarios or experiments. They may also allow an assessment of the incentives to be implemented, in which case the criteria should be used as indicators for the follow-up and efficiency of these incentives.

The number and complexity of tools mean that we should rethink and inventory all questions relative to sustainability and should not aim to free ourselves from political and social arbitration, but on the contrary shed light on it.

A new look at the controversies

A comparison of nutritional and environmental qualities that leads to a reformulation of research questions on animal products and frugality

The *controversy concerning the environmental and nutritional impact of the rise in the share of animal products in diets* is more complex than it first appears, as shown by the case study described in Chapter 3 (even if this study only focused on GHGs, whereas other environmental indicators, such as quantitative and qualitative impacts on water resources and land use, should also be considered). This study, based on actual diets classified according to their compliance with nutritional guidelines, showed that nutritionally adequate diets are characterised by a predominance of plant-based foods, such as staple foods, fruit and vegetables, which generate lower GHG emissions per kilogram than animal products; however, these nutritionally adapted diets contain large quantities of these lower impact foods, so ultimately they are associated with equally high greenhouse gas emissions (in the case of men) or even higher emissions (in the case of women) than nutritionally inadequate diets. *The main point suggested by the study is that reducing the GHG emissions associated with foods will depend more on a reduction in the total quantities ingested than on changes to the structure of food consumption.* If this observation were to be confirmed by future studies, it would undermine some viewpoints in current debates and some of the literature that does not take account of this quantity factor. It poses the question of the conditions for, and feasibility of, an economic model that might *limit the quantities consumed*: to what degree are economic stakeholders in the food chain (via a new economic model for value creation) and consumers prepared for such changes, under what conditions and within what timeframe?

Furthermore, GHG emissions depend on the type of animal products and how they are produced: emissions differ markedly between cattle raised under extensive or intensive conditions, poultry and pig farms. To what extent will improvements in environmental performance on the supply side (from the production of raw materials to the finished product) help to limit GHG emissions and thus reduce the need for profound changes to consumption patterns?

An approach that questions the consideration of upstream sectors

The duALIne project also demonstrated the *merit of specifically studying downstream food sectors, but also the limitations of not including upstream sectors.* GHG emissions downstream of farms are little understood but are estimated to be at least as high as those seen upstream. Thus far, attention has tended to focus upstream; i.e. on agricultural production. Alternatives centred on the concept of agro-ecology do indeed exist, but have no equivalent in the downstream food system, as the use (or production) of ecosystem services at the food processing/ distribution stage is, a priori, limited. Generally speaking, the research questions are not as far advanced: the scientific landscape is fragmented and divided, probably because it involves numerous communities with more varied disciplines than those related to agriculture.

In the food sector, the proportion of budgets devoted to research and development in 2003 (0.25% of sales by European food firms) represented about one-third of the funds set aside for marketing. This low ratio, when compared with other sectors, and notably upstream agriculture (seeds, plant health products and farm equipment) partly explains why this sector is lagging behind in terms of innovation for greater sustainability.

Which conceptual framework for research methods?

Whichever approach is adopted, whether impact assessment methods, system analysis or the study of chain location, studies have demonstrated the importance of temporal and spatial scales and the need to take account of trend dynamics in order to understand challenges and stakeholder behaviours. This conclusion is explained further below.

Taking account of interconnections and combinations

The duALIne project has shown the importance of interconnections and combinations between different dimensions at different levels, for example, combinations of foods in different diets, or the organisation of these diets in time and space. This combination is also observed in food systems that co-exist, interact and evolve over time; it also takes the form of links between different chains and stakeholders of the system governance at different scales, and for different political, economic and social challenges. *This observation questions the research methodologies used to understand all the complexities of these challenges, without considering them separately.* This complexity of combinations and juxtapositions also raises questions concerning the governance of changes and transitions in food systems, which in turn requires the production of specific concepts and tools.

Furthermore, only this type of approach will enable analysis of the issue of system diversity, in terms of behaviour and sectoral organisation, as a source of sustainability or at least of resilience.

Scales

Questions linked to the system's scales of governance need to be studied in depth. At which scales should the sustainability of a system be determined?

- Should it be at the local scale (each firm, each chain or each region)? This would imply efforts to ensure innovation, investment and competitiveness.
- Or should more global sustainability be sought? The greatest efforts should then be made where there is most leeway, even if this means accepting that fewer efforts will be made where there is less leeway.

Within this analytical framework, it would be useful to determine the extent to which such an approach will compensate at larger scales for the positive and negative externalities of the local scale.

Frontiers have been defined for each level of analysis. We need to ensure that any analysis of sustainability will be robust relative to each of these definitions. This implies, on the one hand, that when attention focuses on a particular level, account must be taken of what lies outside this level, and on the other hand that nested levels of the analysis must be linked with the possibilities of inter-level compensation.

How can we understand and analyse the sustainability of food systems that are both intertwined in complex ecosystems and are themselves composite? The problem resides in the dual vision that is necessary, both human-centred and nature-centred.

How can we also take account of temporal scales in these analyses, so that current undertakings will not compromise future sustainability? Combining temporal dynamics (year for production, week for processing, day for distribution, with adaptable storage functions) is a considerable challenge.

A need for organisation

Throughout this analysis, and in light of these final questions, we have seen that it will be necessary to adopt a resolutely cross-disciplinary approach to these questions and provide coordination for the scientific community in order to prevent any fragmentation of the approaches used.

A need for infrastructures

The development of this future research, and of simulation and experimental studies, will all require the availability of large quantities of data and the systems necessary to generate them, which are currently deficient: data-bases on products are needed on the unit stages of processing and on the *domestic practices* of consumers. There is also a lack of longitudinal quantitative analyses of consumption in Southern countries, and of precise quantifications of losses and wastage and of their causes. Repeated, large-scale consumption studies that include vulnerable populations in both developed and developing countries, are necessary. They should collect data that will enable combined study of both the health and environmental dimensions (as a minimum). These studies should be harmonised at an international level, in terms of both the methods used and product identification.

Viewpoints to be explored, gaps to be filled

This section covers all the issues that were not addressed by the duALIne project and which therefore require bibliographical analysis and the formula-tion of questions for research.

Poverty and food security

Reducing social inequalities between and within countries is no longer seen as a purely moral issue. Economic insecurity and social marginalisation generate costs for society in terms of health, as poor populations are more vulnerable, notably to nutritional risks. These problems increase the risks of social instability (e.g. hunger riots), the development of parallel economies (or even illegal crop growing) and environmental negligence. Poverty is therefore both a consequence and cause of a non-sustainable economic system.

Even if we are able to add environmental, social, health or nutritional attrib-utes to food, does the same also apply to a reduction in social inequalities? Fair trade labelling is designed to ensure better prices for poor agricultural producers. But it does not solve the problem of poor consumers. The prospect of food supplies dedicated to vulnerable populations, in the form of aids, grants or products designed according to their needs, may compensate for an economic model that generates insecurity without tackling the process of pauperisation.

Although the social pillar is theoretically included in projects on sustainable development, in practice it is often neglected in favour of the environmental and economic pillars. Research efforts are necessary to produce measurement tools and references on the effects of changes to the food system on inequalities and poverty. We also need to better characterise the resources and drivers that can be activated to reduce the vulnerability of precarious populations and ensure their better inclusion in society. This supposes both identifying and understanding how they achieve resilience, and testing and assessing interventions and policies designed to reduce inequalities.

Governance issues

One question concerns the respective roles of the different stakeholders concerned: governments, the private sector, local government bodies and NGOs. Whether concerning the environment, health or social equity, we are now seeing a considerable focus on voluntary standards of non-governmental stakeholders, private enterprises or NGOs. The governance of sustainable development is no longer the preserve of governments. Indeed, in many Southern countries, the governments do not even have the technical, human, legal or financial resources necessary to intervene. What is the efficiency of these new types of governance in attaining the objectives they are meant to be targeting? Is it solely a question of advertising, or are they really capable of driving changes in practices?

Another question concerns the *participative mode* of this type of governance. The involvement of the stakeholders concerned, and particularly of representatives from civilian society and the private sector, in the definition of interventions at both the local and global scales, can tend to be somewhat problematic. It is often deemed virtuous, but has also proved ambivalent: on the one hand, it is intended to avoid conflict, and on the other, it reduces the possibility of measuring the balance of power or of initiating a true public debate; in other words, of addressing political questions. In this sense, we can refer to the risks of depoliticising public actions.

Other questions not covered

Water and biodiversity must be included in impact analyses. They also generate other research questions.

Water

The issue of water is mentioned in Chapters 6 and 10. Although its management is crucial to productivity and to different sectors, both as a resource and as a constitutive element or ingredient in products, too few indicators take it into account. Its quantitative and qualitative availability must be integrated in any analysis of spatial dynamics and stakeholder strategies in a given system; for example, in regions where it has been decided to develop both food industries and market gardening, the sustainable management of water resources will be a determining factor. It can be seen that at a local scale, water may become a crucial factor in decision-making, and a high priority input variable. In the anticipated context of depleted water resources, studies on water, which has not been considered in detail in duALIne, need to be pursued.

Biodiversity

The links between eating patterns and biodiversity, although often mentioned in debates on the future of food, are in fact poorly documented in the scientific literature, whether in terms of diversity within and between varieties, or that of cultivated or wild species linked to cropping systems, or the diversity of ecosystems and rural landscapes. In the same way, little work has been done on measuring biodiversity and characterising its relationships with food systems, as is mentioned in Chapters 4 and 10. Much work still needs to be done to understand how food systems could be designed (from upstream to downstream) to maintain ecosystem services and biodiversity and to determine their maintenance threshold.

Food safety

The issue of food safety in Southern countries, referred to in Chapter 4 but not directly addressed during the duALIne project, has not been solved in these countries and remains on the research agenda. This poses the question of the universal or relative nature of safety standards in markedly different contexts.

Other questions were not addressed during duALIne, and should be studied in terms of their links with food sustainability: product traceability and the social control of technological innovations such as nanotechnologies or biotechnologies.

What are the future prospects?

A problem that is not necessarily shared with Southern countries

The concept of food sustainability originated from increased awareness of the limits of the dominant agri-industrial food system with respect to its impacts on the planet. Today, it is becoming a recurrent issue in industrialised countries. But because this system is tending to spread throughout the planet, it is tempting to situate this debate at a global scale, and thus include Southern countries. However, in the latter case, the problems are more numerous: even if these countries are now starting to face the same problems of sustainability previously seen in industrialised countries – and notably the rapid development of obesity in part of the population – they still need to manage the problems linked to recent or persistent under-nutrition: in the agri-food sector, the issues of nutrient efficiencies, post-harvest losses, food safety, labour productivity and energy performance are far from being resolved. So these questions should not be set aside in research strategies and work should not focus only on the new questions of sustainability posed in more industrialised countries. Furthermore, these questions are not necessarily a concern unanimously shared by countries which consider that efforts to limit environmental, social or safety damage are first of all incumbent upon those who have caused them. Public pressure in favour of systems that are better for the environment, health or equity is – still – relatively weak in Southern countries, when compared to the situation in the most industrialised countries.

What can be deduced from this?

We need to be able to establish dialogue with stakeholders from Southern countries on the prospects of trend changes in the food system. This supposes considering questions of long-term trends as they are seen by these countries. We must reinforce analytical capacities in these countries, where they are currently limited, so that they can assess and determine their policies in an informed manner. Their food systems are changing at a very rapid rate, and there is a risk that this situation may generate irreversible damage. However, these countries are not the only stakeholders in these evolutions because food systems are becoming increasingly interconnected.

Research to support debate and decision-making in a context of growing uncertainty

Because of the growing uncertainty of the context, and the complexity of the systems and methods used to assess their impacts, and in view of the multiplicity of stakeholders concerned, there will be fewer and fewer clear recommendations for these stakeholders. The question will no longer simply be to justify for the measures implemented, but to try and determine how to govern in a situation of growing uncertainty without fully knowing the consequences of a particular decision or choice over another.

In this context, we feel it is necessary to initiate a foresight study on food so as to shed light on the areas of uncertainty concerning the possible trends that will affect food systems, and their consequences. This study will complement *Agrimonde* (Paillard *et al.*, 2010) by focusing on downstream aspects of the systems, and particularly on changes to consumption.

Trade-offs will need to be made between the challenges of sustainable development, because it is highly probable that it will not be possible to determine an option that can win on all fronts at the same time. The role of research will be to supply the tools that can describe these challenges and the bases for decision-making.

The duALIne project has proposed a panorama of *questions for research* identified through a review of the scientific literature and backed by specific studies conducted within a framework of academic rigour. Among all these questions, the priorities need to be determined as part of a political process, in the primary sense of the term, i.e. which concerns all stakeholders in the city (*polis*).

Nevertheless, some important points and major needs have emerged from this exercise and are described above. In our opinion, they should be placed on the research agenda as soon as possible. Their confrontation with current research fronts will determine how they are taken into account by research teams.

For INRA, these issues will be included in its new strategy for the creation of meta-programmes; they could be addressed by the programme on dietary behaviour and by the programme focusing on world food security and global change.

For CIRAD, the results of this project will contribute to updating its scientific strategy, particularly concerning its research areas on 'Accessible, diversified and safe food', and 'Public policies, poverty and inequalities'.

International dissemination

The research questions and infrastructure needs highlighted by duALIne will be picked up and debated at the *international level*, notably in the context of the FACCE Joint Programming Initiative (Agriculture, Food Security and Climate Change) and the SUSFOOD (Sustainable Food) ERA-Net, a European research network on the theme of food sustainability. They will also be the subject of discussions with Southern countries in the context of governance bodies focused on global food security.

References

Abbott, P. 2010. Stabilisation policies in developing countries after the 2007–2008 food crisis. *52. Session of the Working Party on Agricultural Policy and* Markets, 2010/11/15–17. Paris: OECD, 54 pp. Available at: http://www.oecd.org/dataoecd/50/34/46340396.pdf.

ADEME 2007. *Campagne nationale de caractérisation des ordures ménagères 2007*. Paris: Agence de l'environnement et de la maîtrise d'énergie. Available at: http://www2.ademe.fr/servlet/KBaseShow?sort=-1&cid=96&m=3&catid=15163.

ADEME 2010. *Guide des facteurs d'émission*. Paris: ADEME, 240 pp.

Adger, W. N. 2003. Social capital, collective action, and adaptation to climate change. *Economic Geography*, 79 (4): 387–404.

Adger, W. N., Arnell, N. W. and Tompkins, E. L. 2005. Successful adaptation to climate change across scales. *Global Environmental Change-Human and Policy Dimensions*, 15 (2): 77–86.

Adoue, C. 2007. *Mettre en oeuvre l'écologie industrielle*. Lausanne, Switzerland: Presses polytechniques et universitaires romandes, 128 pp.

AFD-FFEM 2010. *Indications géographiques: qualité des produits, environnement et cultures*. Paris: AFD (Agence française de Développement) – Fonds français pour l'Environnement mondial (FFEM) (*Savoirs Communs*), pp. 1–100. Available at: http://www.afd.fr/jahia/webdav/site/afd/shared/PUBLICATIONS/THEMATIQUES/savoirscommuns/09-Savoirs-communs.pdf.

AFNOR 2006a. ISO 14040:2006. Management environnemental: Analyse du cycle de vie – Principes et cadre. Paris: AFNOR, 23 pp.

AFNOR 2006b. ISO 14044:2006. Management environnemental: Analyse du cycle de vie – Exigences et lignes directrices. Paris: AFNOR, 49 pp.

Ajslev, T. A., Andersen, C. S., Gamborg, M., Sorensen, T. I. A. and Jess, T. 2011. Childhood overweight after establishment of the gut microbiota: the role of delivery mode, pre-pregnancy weight and early administration of antibiotics. *International Journal of Obesity*, 35 (4): 522–529. Available at: http://dx.doi.org/10.1038/ijo.2011.27.

Alkire, S. 2010. Multidimensional poverty and its discontents. AFD-EUDN Conference, Paris, 28 pp. Available at: http://www.ophi.org.uk/wp-content/uploads/OPHI-RP-23a.pdf.

Alkire, S. and Santos, M. E. 2010. Acute Multidimensional Poverty: A New Index for Developing Countries, OPHI Working Paper, 38. Oxford: Oxford Poverty and Human Development Initiative, 139 pp. Available at: http://www.ophi.org.uk/wp-content/uploads/ophi-wp38.pdf.

Almond, D., Edlund, L., Li, H. and Zhang, J. 2010. Long-term effects of early-life development: evidence from the 1959 to 1961 China famine. In T. Ito and A. Rose, eds., *The Economic Consequences of Demographic Change in East Asia*. Chicago, IL: University of Chicago Press, pp. 321–345. Available at http://www.nber.org/chapters/c8166.

Alston, J. M., Sumner, D. A. and Vosti, S. A. 2006. Are agricultural policies making us fat? Likely links between agricultural policies and human nutrition and obesity, and their policy implications. *Applied Economic Perspectives and Policy*, 28 (3): 313–322.

Alston, J. M., Sumner, D. A. and Vosti, S. A. 2008. Farm subsidies and obesity in the United States: national evidence and international comparisons. *Food Policy*, 33 (6): 470–479. Available at: http://dx.doi.org/10.1016/j.foodpol.2008.05.008.

Anastas, P. and Eghbali, N. 2010. Green chemistry: principles and practice. *Chemical Society Reviews*, 39 (1): 301–312. Available at: http://dx.doi.org/10.1039/B918763B.

Anderson, K. and Nelgen, S. 2012. Trade barrier volatility and agricultural price stabilization. *World Development*, 40: 36–48. Available at: http://www.sciencedirect.com/science/article/pii/S0305750X11001483.

Anderson, P. C., Cummings, L., Schippers, T. and Simonel, B. 2002. *Le traitement des récoltes: un regard sur la diversité, du néolithique au présent. Actes des XXIIIèmes rencontres internationales d'archéologie et d'histoire d'Antibes, 17–19 October 2002, Antibes, 2003*. Antibes: Editions APDCA, 524 pp.

Anderson, S., Allen, J. and Browne, M. 2005. Urban logistics: how can it meet policy makers' sustainability objectives? *Journal of Transport Geography*, 13 (1): 71–81. Available at: http://dx.doi.org/10.1016/j.jtrangeo.2004.11.002.

Aounallah-Skhiri, H., Traissac, P., El Ati, J. *et al.* 2011. Nutrition transition among adolescents of a south-Mediterranean country: dietary patterns, association with socioeconomic factors, overweight and blood pressure. A cross-sectional study in Tunisia. *Nutrition Journal*, 10 (1): 38. Available at: http://dx.doi.org/doi:10.1186/1475-2891-10-38.

Apfelbaum, M. 1998. *Risques et peurs alimentaires*. Paris: Odile Jacob, 284 pp.

Appadurai, A. 1986. The social life of things: commodities in cultural perspective. Ethnohistory Workshop, 1983–84: Symposium on the Relationship between Commodities and Culture, 23–25 May. Philadelphia. Cambridge and New York: Cambridge University Press, 329 pp.

Arnoult, M. H., Jones, P. J., Tranter, R. B. *et al.* 2010. Modelling the likely impact of healthy eating guidelines on agricultural production and land use in England and Wales. *Land Use Policy*, 27 (4): 1046–1055. Available at: http://dx.doi.org/10.1016/j.landusepol.2010.02.001.

ARP ADAGE. ADaptation de l'Agriculture et des Ecosystèmes anthropisés au changement climatique. Available at: http://www1.clermont.inra.fr/adage/, accessed 11 July 2011.

Arumugam, M., Raes, J., Pelletier, E. *et al.* 2011. Enterotypes of the human gut microbiome. *Nature*, 474 (7353): 666. Available at: http://dx.doi.org/10.1038/nature10187.

Ascher, F. 2005. *Le mangeur hypermoderne: une figure de l'individu éclectique*. Paris: Odile Jacob, 330 pp.

Atkinson, A. B. and Marlier, E. 2010. *Income and Living Conditions in Europe*. Luxembourg: European Commission. 424 pp. Available at: http://epp.eurostat.ec.europa.eu/cache/ITY_OFFPUB/KS-31-10-555/EN/KS-31-10-555-EN.PDF.

Aubin, J. P. 1991. *Variability Theory*. Boston, MA: Birkhäuser, 543 pp.

Ayres, R. U. 2000. Commentary on the utility of the ecological footprint concept. *Ecological Economics*, 32 (3): 347–349. Available at: http://dx.doi.org/10.1016/S0921-8009(99)00151-2.

Bagoulla, C., Chevassus-Lozza, E., Daniel, K. and Gaigné, C. 2010. Regional production adjustment to import competition: evidence from the French agro-industry. *American Journal of Agricultural Economics*, 92 (4): 1040–1050. Available at: http://dx.doi.org/10.1093/ajae/aaq053.

Bairoch, P. 1997. Victoires et déboires: histoire économique et sociale du monde du XVIe siècle à nos jours. Collection Folio. Histoire n°78–80. Paris: Gallimard, 662 pp.

Balasubramanian, P. R. and Bai, R. K. 1992. Recycling of biogas-plant effluent through aquatic plant (*Lemna*) culture. *Bioresource Technology*, 41 (3): 213–216. Available at: http://dx.doi.org/10.1016/0960-8524(92)90004-h.

Barbosa-Canovas, G. V., Mortimer, A., Lineback, D. *et al.* 2009. *Global Issues in Food Science and Technology*. Amsterdam, The Netherlands and Burlington, MA: Elsevier and Academic Press, 520 pp.

Baributsa, D., Lowenberg-DeBoer, J., Murdock, L. and Moussa, B. 2010. Profitable chemical-free cowpea storage technology for smallholder farmers in Africa: opportunities, and challenges. 10th International Working Conference on Stored Product Protection. 27 June to 2 July 2010, Estoril, Portugal. Julius-Kühn-Archiv, 1046–1052. Available at: http://pub.jki.bund.de/index.php/JKA/article/download/1319/1364.

Baris, P. and Cordier, J. 2011. Le risque prix sur les produits alimentaires importés: outils de couverture pour l'Afrique. Paris: Agence française de développement (A savoir, n°10), 95 pp.

Barker, D. J. P. 1998. *Mothers, Babies, and Health in Later Life*. Edinburgh: Churchill Livingstone, 217 pp. Available at: http://books.google.com/books/about/Mothers_babies_and_health_in_later_life.html?id=HtfY0furyrUC.

Barnett, A., Pyle, L. and Subramanian, S. K. 1978. *Biogas Technology in the Third World: A Multidisciplinary Review*. Ottawa, Canada: International Development Research Centre, 132 pp. Available at: http://idl-bnc.idrc.ca/dspace/bitstream/10625/3493/1/29731.pdf.

Baroni, L., Cenci, L., Tettamanti, M. and Berati, M. 2006. Evaluating the environmental impact of various dietary patterns combined with different food production systems. *European Journal of Clinical Nutrition*, 61 (2): 279–286. Available at: http://dx.doi.org/10.1038/sj.ejcn.1602522.

Barrett, C. B. 2010. Measuring food insecurity. *Science*, 327 (5967): 825–828.

Barrientos, S., Dolan, C. and Tallontire, A. 2001. Gender and ethical trade: a mapping of the issues in African horticulture. Project No. R7525, 41 pp. Available at: http://www.nri.org/projects/nret/genderet.pdf.

Basset-Mens, C. and van der Werf, H. 2004. Evaluation environnementale de systèmes de production de porc contrastés. *36èmes Journées de la Recherche Porcine:* 2004/02/03–05. Paris: Institut Technique du Porc, pp. 165–172. Available at: http://www.journees-recherche-porcine.com/texte/index.htm.

Bebbington, A. 1999. Capitals and capabilities: a framework for analyzing peasant viability, rural livelihoods and poverty. *World Development*, 27 (12): 2021–2044. Available at: http://dx.doi.org/10.1016/S0305-750X(99)00104-7.

Bellon, S., Gautronneau, Y., Riba, G. *et al.* 2000. L'agriculture biologique et l'INRA: vers un programme de recherche. *INRA Mensuel* (sup 104): 1–25. Available at: http://www.inra.fr/internet/Directions/DIC/ACTUALITES/Agribio/AgribioSom.htm.

Berdegué, J. A., Balsevich, F., Flores, L. and Reardon, T. 2005. Central American supermarkets' private standards of quality and safety in procurement of fresh fruits and vegetables. *Food Policy*, 30 (3): 254–269. Available at: http://dx.doi.org/10.1016/j.foodpol.2005.05.003.

Bere, E. and Brug, J. 2009. Towards health-promoting and environmentally friendly regional diets? A Nordic example. *Public Health Nutrition*, 12 (01): 91–96. Available at: http://dx.doi.org/10.1017/S1368980008001985.

Bio Intelligence Service 2010. Preparatory study on food waste across EU 27. Final report. Contract no. 07.0307/2009/540024/SER/G4 European Commission (DG ENV) Directorate C – Industry, 210 pp. Available at: http://ec.europa.eu/environment/eussd/pdf/bio_foodwaste_report.pdf.

Blackman, A. and Rivera, J. 2010. The evidence base for environmental and socioeconomic impacts of 'sustainable' certification. Discussion paper. Washington: Resources for the Future, 31 pp. Available at: http://www.rff.org/rff/documents/RFF-DP-10-17.pdf.

Blanford, D. 1984. Changes in food consumption patterns in the OECD area. *European Review of Agricultural Economics*, 11 (1): 43–64. Available at: http://dx.doi.org/10.1093/erae/11.1.43.

Blas, J. 2009. Food security fuels land grab, says report. *Financial Times*, 24 May. Available at: http://farmlandgrab.org/3043.

Bockstaller, C., Espagnol, S., Guichard, L., Petit, J., Raison, C. and Vertés, F. 2010. Synthèse 1. Stratégies de choix des méthodes et outils d'évaluation environnementale en systèmes d'élevage. In S. Espagnol and P. Leterme, eds., *Élevages et Environnement*. Versailles, France: Quae (Sciences en partage), pp. 15–65.

Bockstaller, C., Guichard, L., Makowski, D. *et al.* 2008. Agri-environmental indicators to assess cropping and farming systems. A review. *Agronomy for Sustainable*

Development, 28 (1): 139–149. Available at: http://dx.doi.org/10.1051/
agro:2007052.

Bohanec, M., Zupan, B. and Rajkovic, V. 2000. Applications of qualitative
multi-attribute decision models in health care. *Journal of Medical Informatics*,
58–59: 191–205. Available at: http://dx.doi.org/10.1016/S1386-5056(00)00087-3.

Boiffin, J., Hubert, B. and Durand, N. 2004. *Agriculture et développement durable. Enjeux et
questions de recherche*. Paris (FRA): INRA Editions, 91 pp. Available at: http://www.
inra.fr/developpement-durable/agriculture-developpement.pdf.

Boltanski, L. and Chiapello, E. 1999. *Le nouvel esprit du capitalisme*. Paris: Gallimard,
843 pp.

Bonnet, C. and Réquillart, V. 2010. Does the EU sugar policy reform increase added
sugar consumption? An empirical evidence on the soft drink market. TSE
Working Papers (No.10–197), Toulouse School of Economics (TSE), France 18 pp.
Available at: http://ideas.repec.org/p/tse/wpaper/23467.html.

Bontemps, C., Maigne, E. and Réquillart, V. 2013 La productivité de l'agro-alimentaire
français de 1996 à 2006. *Economie et Prévision* (in press).

Boserup, E. 1975. The impact of population growth on agricultural output. *The
Quarterly Journal of Economics*, 89 (2): 257–270. Available at: http://dx.doi.org/
10.2307/1884430.

Boström, M. and Klintman, M. 2008. *Eco-standards, Product Labelling and Green
Consumerism*. New York: Palgrave Macmillan, 256 pp.

Bouly de Lesdain, S. 2002. Alimentation et migration, une définition spatiale. In
I. Garabuau-Moussaoui, E. Palomares and D. Desjeux, eds, *Alimentations
Contemporaines*. Paris: L'Harmattan, pp. 173–189. Available at: http://halshs.
archives-ouvertes.fr/docs/00/12/07/62/PDF/Alimentation_et_migration.pdf.

Boutaud, A. and Gondran, N. 2009. *L'empreinte écologique*. Paris: La Découverte
(collection Repères), 122 pp.

Bove, C. F. and Sobal, J. 2006. Foodwork in newly married couples: making family
meals. *Food, Culture and Society: An International Journal of Multidisciplinary
Research*, 9: 69–89. Available at: http://dx.doi.org/10.2752/
155280106778055118.

Bowen, S. and Valenzuela Zapata, A. 2008. Designations of origin and socioeconomic
and ecological sustainability: the case of tequila in Mexico. *Cahiers Agricultures*, 17
(6): 552–560. Available at: http://dx.doi.org/10.1684/agr.2008.0241.

Branger, A., Richer, M.-M. and Roustel, S. 2007. *Alimentation et processus technologiques*.
Dijon, France: Educagri, 293 pp.

Braudel, F. 1979. *Civilisation matérielle, économie et capitalisme, 15ème-18ème siècle*. Paris:
Armand Colin.

Bray, D. B., Sanchez, J. L. P. and Murphy, E. C. 2002. Social dimensions of organic
coffee production in Mexico: lessons for eco-labeling initiatives. *Society & Natural
Resources*, 15 (5): 429–446. Available at: http://dx.doi.org/10.1080/
08941920252866783.

Brenton, P., Edwards-Jones, G. and Jensen, M. F. 2009. Carbon labelling and
low-income country exports: a review of the development issues. *Development*

Policy Review, 27 (3): 243–267. Available at: http://www.oecd.org/dataoecd/47/21/42930328.pdf.

Bricas, N. and Seck, P. A. 2004. L'alimentation des villes du Sud: les raisons de craindre et d'espérer. *Cahiers Agriculture*, 13 (1): 10–14. Available at: http://www.john-libbey-eurotext.fr/fr/revues/agro_biotech/agr/e-docs/00/03/FE/DD.

Bricas, N., Thirion, M.-C. and Zoungrana, B. 2009. Bassins de production et de consommation des cultures vivrières en Afrique de l'Ouest et du Centre. Rapport provisoire: CIRAD, AFD, CILSS, 51 pp. Available at: http://www.cilss.bf/IMG/pdf/Bassins_vivriers_Resoultion_Moyenne_.pdf.

Brondel, L., Romer, M., Van Wymelbeke, V. *et al.* 2009. Variety enhances food intake in humans: role of sensory-specific satiety. *Physiology & Behavior*, 97 (1): 44–51. Available at: http://dx.doi.org/10.1016/j.physbeh.2009.01.019.

Brown, C. and Miller, S. 2008. The impacts of local markets: a review of research on farmers markets and community supported agriculture (CSA). *American Journal of Agricultural Economics*, 90 (5): 1298–1302. Available at: http://ajae.oxfordjournals.org/content/90/5/1298.short.

Brown, L. R. 2011. *World on the Edge: How to Prevent Environmental and Economic Collapse.* New York: W.W. Norton, 240 pp.

Bruinsma, J. 2003. *World Agriculture: Towards 2015/2030: An FAO Perspective.* London: Earthscan Publications Ltd, 432 pp. Available at: http://www.fao.org/fileadmin/user_upload/esag/docs/y4252e.pdf.

Bruinsma, J. 2009. The resource outlook to 2050: by how much do land, water and crop yields need to increase by 2050? *How to Feed the World in 2050. Proceedings of a Technical Meeting of Experts, Rome, Italy, 24–26 June 2009*, pp. 1–33. Available at: http://www.fao.org/docrep/012/ak542e/ak542e00.htm.

Brundtland, G. H. 1987. *Our Common Future.* Oxford: Oxford University Press and the United Nations World Commission on Environment and Development, 400 pp.

Butault, J.-P. 2008. La relation entre prix agricoles et prix alimentaires. *Revue Française d'Economie*, 23 (23–2): 215–241. Available at: http://www.persee.fr/articleAsPDF/rfeco_0769-0479_2008_num_23_2_1670/article_rfeco_0769-0479_2008_num_23_2_1670.pdf?mode=light.

Butland, B., Jebb, S., Kopelman, P. *et al.* 2008. *Tackling Obesities: Future Choices. Summary of Key Messages.* London: Government Office for Science. 4 pp. Available at: http://www.bis.gov.uk/assets/bispartners/foresight/docs/obesity/20.pdf.

Byerlee, D., Jayne, T. S. and Myers, R. J. 2006. Managing food price risks and instability in a liberalizing market environment: overview and policy options. *Food Policy*, 31 (4): 275–287. Available at: http://www.sciencedirect.com/science/article/pii/S030691920600025X.

Caillavet, F., Lecogne, C. and Nichèle, V. 2009. La consommation alimentaire: des inégalités persistantes mais qui se réduisent. In *Cinquante ans de consommation en France*. Paris: INSEE, pp. 49–62. Available at: http://insee.fr/fr/ffc/docs_ffc/ref/CONSO09e.PDF.

Calvo, E. 1997. Toujours Africains et déjà Français: la socialisation des migrants vue à travers leur alimentation. *Politique Africaine*, 67 (3): 48–55. Available at: http://www.politique-africaine.com/numeros/pdf/067048.pdf.

Capitaine, M., David, C. and Freycenon, R. 2009. Evaluation et amélioration de la durabilité de l'agriculture biologique: éléments de débats. *Innovations Agronomiques*, 4: 209–215. Available at: http://www.inra.fr/ciag/content/download/3284/29926/version/1/file/30-Capitaine.pdf.

Capt, D. 2008. Permanence et changement dans la production et la commercialisation des produits fermiers. Une approche économique. *Séminaire INRA AgroParisTech: Filières courtes de distribution, développement durable et territoires*. Paris: 2008/04/08. Oral communication, 33 pp. Available at: http://www.psdr-coxinel.fr/IMG/pdf/4_INRA_SAD-APT_seminaire_Filieres_Courts_DD_et_Territoire_DCapt.pdf.

Capt, D. and Wavresky, P. 2010. Evolution des exploitations agricoles françaises en vente directe et circuits. *Colloque National Circuits Courts alimentaires*, 2010/05/05–06. Paris: AgroParisTech.

Carlsson-Kanyama, A., Ekström, M. P. and Shanahan, H. 2003. Food and life cycle energy inputs: consequences of diet and ways to increase efficiency. *Ecological Economics*, 44 (2–3): 293–307. Available at: http://dx.doi.org/10.1016/S0921-8009(02)00261-6.

Carlsson-Kanyama, A. and Gonzalez, A. D. 2009. Potential contributions of food consumption patterns to climate change. *American Journal of Clinical Nutrition*, 89 (5): S1704–S1709. Available at: http://dx.doi.org/10.1038/sj.ejcn.1602522.

Carpenter, S. R., Pingali, P. L., Bennett, E. M. and Zurek, M. B. 2005. *Ecosystems and Human Well-being: Scenarios, Vol. 2*. Washington DC: Island Press, 560 pp. Available at: http://www.millenniumassessment.org/en/Scenarios.html.

Carpio, C. E. and Isengildina-Massa, O. 2009. Consumer willingness to pay for locally grown products: the case of South Carolina. *Agribusiness*, 25 (3): 412–426. Available at: http://dx.doi.org/10.1002/agr.20210.

CAS 2009. *Approche économique de la biodiversité et des services liés aux écosystèmes: contribution à la décision publique*. Paris: La Documentation française (Rapports et documents), 399 pp.

Cederberg, C. and Stadig, M. 2003. System expansion and allocation in life cycle assessment of milk and beef production. *International Journal of Life Cycle Assessment*, 8 (6): 350–356. Available at: http://dx.doi.org/10.1065/lca2003.07.126.

Cépède, M. and Lengellé, M. 1953. *Économie alimentaire du globe; essai d'interprétation*. Paris: Librairie de Médicis, M.-T. Génin, 654 pp.

Cépède, M. and Lengellé, M. 1970. *L'économie de l'alimentation*. Paris: PUF, 123 pp.

Chantrel, E. and Lecocq, P. E. 2009. Les marges dans la filière agro-alimentaire en France. *Economie et Prévision*, 189: 141–149.

Chapoto, A. and Jayne, T. 2010. Do market interventions promote food price stability? Evidence from Eastern and Southern Africa. *Awakening the Sleeping Giant: Making Grain Markets Work for Smallholder Farmers and Consumers in Eastern and Southern Africa*. Lusaka Zambia. 10 May, 2010. ACTESA/COMEA. Available at: http://www.aec.msu.edu/fs2/gisama/Session_2-Effects_Market_Interventions_on_Price_Stability-Chapoto.pdf.

Charles, M.-A., Eschwege, E. and Basdevant, A. 2008. Monitoring the obesity epidemic in France: The Obepi Surveys 1997–2006. *Obesity*, 16 (9): 2182–2186. Available at: http://dx.doi.org/10.1038/oby.2008.285.

Chen, Y. and Zhou, L.-A. 2007. The long-term health and economic consequences of the 1959–1961 famine in China. *Journal of Health Economics*, 26 (4): 659–681. Available at: http://www.sciencedirect.com/science/article/B6V8K-4MSXT4D-1/2/a74525191177a83610609d9e5733d4bb.

Chevassus-au-Louis, B. 2002. Les crises alimentaires sont-elles durables? In Collectif, ed., *Prévision, analyse et gestion du risque alimentaire (14°èms rencontres scientifiques et technologiques des industries alimentaires, Nancy, 2002)*. Paris: Lavoisier, pp. 21–38.

CIHEAM 2008. *Mediterra 2008: The Future of Agriculture and Food in Mediterranean Countries*. Paris: Presses de Sciences Po, 360 pp. Available at: http://www.ciheam.org/images/CIHEAM/PDFs/Publications/Mediterra/2008/MEDITERRA08FR.pdf.

CNE. 2011. Prévention du gaspillage et des pertes des produits de grande consommation: le rôle clé de l'emballage. Paris: Conseil national de l'emballage, 19 pp. Available at: http://www.conseil-emballage.org/img/publications/74_1.pdf.

Cohen, L. 2003. *A Consumers' Republic: The Politics of Mass Consumption in Postwar America*. New York: Alfred A. Knopf, 567 pp.

Coley, D., Howard, M. and Winter, M. 2009. Local food, food miles and carbon emissions: a comparison of farm shop and mass distribution approaches. *Food Policy*, 34 (2): 150–155. Available at: http://dx.doi.org/10.1016/j.foodpol.2008.11.001.

Coley, D. A., Goodliffe, E. and Acdiarmid, J. 1998. The embodied energy of food: the role of diet: viewpoint. *Energy Policy*, 26 (6): 455–459. Available at: http://dx.doi.org/10.1016/S0301-4215(97)00159-6.

Collier, P. 2007. *The Bottom Billion: Why the Poorest Countries are Failing and What Can be Done About it*. Oxford: Oxford University Press, 205 pp. Available at: http://www.oup.com/us/catalog/general/subject/Economics/Developmental/?view=usa&ci=9780195311457.

Colonna, P. (coord.) 2006. *La chimie verte*. Paris and New York: Éditions Tec. & Doc., 532 pp.

Combes, P.-P. and Lafourcade, M. 2005. Transport costs: measures, determinants, and regional policy implications for France. *Journal of Economic Geography*, 5 (3): 319–349. Available at: http://joeg.oxfordjournals.org/content/5/3/319.abstract.

Combris, P. 2006. Le poids des contraintes économiques dans les choix alimentaires. *Cahiers de Nutrition et de Diététique*, 41 (5): 279–284. Available at: http://www.em-premium.com/article/79185.

Commissariat Général au Développement Durable, 2010. *Les indicateurs de la stratégie nationale de développement durable 2010–2013*. Paris: Commissariat Général au Développement Durable, 48 pp. Available at: http://www.developpement-durable.gouv.fr/IMG/pdf/Les_indicateurs_de_la_strategie_nationale_de_DD_V3.pdf.

Conner, D. 2004. Beyond organic: information provision for sustainable agriculture in a changing market. *Journal of Food Distribution Research*, 35 (1): 34–39. Available at: http://ageconsearch.umn.edu/bitstream/27137/1/35010034.pdf.

Corbeau, J.-P. 2000. Cuisiner, manger, métisser . . . Révolution dans les cuisines. *Revue des Sciences Sociales* (27): 68–73. Available at: http://www.revue-des-sciences-sociales.com/pdf/rss27-corbeau.pdf.

Cotula, L., Vermeulen, S., Leonard, R. and Keeley, J. 2009. *Land Grab or Development Opportunity? Agricultural Investment and International Land Deals in Africa*. London, Rome: FAO, IIED and IFAD, 120 pp. Available at: http://www.fao.org/docrep/011/ak241e/ak241e00.htm.

Coulter, J. 2010. Warehouse receipting, loss reduction, and the development of value chains for grains. In *Reducing Post-harvest Losses in Grain Supply Chains in Africa*. Rome: FAO and The World Bank, pp. 16–18. Available at: http://reliefweb.int/sites/reliefweb.int/files/resources/5506BABD1F27BFC0C12577490047D159-FAO_Mar2010.pdf.

Coulter, J. and Shepherd, A. W. 2001. Le crédit sur nantissement des stocks – Une straté gie de développement des marches agricoles. *Bulletin des services agricoles de la FAO* (120): 69 pp. Available at: ftp://ftp.fao.org/docrep/fao/012/v7470f/v7470f00.pdf.

Coursey, D. G. and Booth, R. H. 1972. The postharvest phytopathology of perishable tropical produce. *Review of Plant Pathology*, 51 (12): 751–765.

Coveney, J. 2005. A qualitative study exploring socio-economic differences in parental lay knowledge of food and health: implications for public health nutrition. *Public Health Nutrition*, 8 (3): 290–297. Available at: http://dx.doi.org/10.1079/PHN2004682.

Cowell, S. J. and Parkinson, S. 2003. Localisation of UK food production: an analysis using land area and energy as indicators. *Agriculture, Ecosystems & Environment*, 94 (2): 221–236. Available at: http://dx.doi.org/10.1016/S0167-8809(02)00024-5.

Craig, W. and Mangels, A. 2009. Position of the American Dietetic Association: vegetarian diets. *Journal of the American Dietetic Association*, 109 (7): 1266–1282. Available at: http://dx.doi.org/10.1016/j.jada.2009.05.027.

Cuellar, A. D. and Webber, M. E. 2010. Wasted food, wasted energy: the embedded energy in food waste in the United States. *Environmental Science & Technology*, 44 (16): 6464–6469. Available at: http://pubs.acs.org/doi/pdfplus/10.1021/es100310d.

Curran, M., de Baan, L. and De Schryver, A. M. *et al.* 2011. Toward meaningful end points of biodiversity in life cycle assessment. *Environmental Science & Technology*, 45 (1): 70–79. Available at: http://dx.doi.org/10.1021/es101444k.

D'Andrea, A. C. 2003. Social and technological aspects of non-mechanised emmer processing. In P. C. Anderson, L. S. Cummings, T. K. Schippers and B. Simonel, eds, *Le Traitement des récoltes: un regard sur la diversité du Néolithique au présent. Actes des Rencontres, 17–19 octobre 2002*. Antibes: Editions APDCA, pp. 41–60.

Da Silva, V. P., Van der Werf, H. M. G. and Soares, S. R. 2010. LCA of French and Brazilian broiler poultry production systems. *VII International Conference on Life Cycle Assessment in the Agri-food Sector*. Bari, 22–24 September 2010, oral communication, n.p.

Darby, K., Batte, M. T., Ernst, S. and Roe, B. 2008. Decomposing local: a conjoint analysis of locally produced foods. *American Journal of Agricultural Economics*, 90 (2): 476–486. Available at: http://ajae.oxfordjournals.org/content/90/2/476.abstract.

Darmon, M. and Briend, A. 2001. Contraintes et coût d'une alimentation équilibrée. *Cuisine Collective* (140): 1–2. Available at: http://www.la-cuisine-collective.fr/archives/2001/avril01/qualite/Contraintesetcout.html.

Darmon, N., Ferguson, E. L. and Briend, A. 2006. Impact of a cost constraint on nutritionally adequate food choices for French women: an analysis by linear programming. *Journal of Nutrition Education and Behavior*, 38 (2): 82–90. Available at: http://dx.doi.org/10.1016/j.jneb.2005.11.028.

Daviron, B. 2002. Small farm production and the standardization of tropical products. *Journal of Agrarian Change*, 2 (2): 162–184. Available at: http://dx.doi.org/10.1111/1471-0366.00029.

de Benoist, B., McLean, E., Egli, I. and Cogswell, M. 2008. *Worldwide Prevalence of Anaemia 1993–2005: WHO Global Database on Anaemia*. Geneva: World Health Organization, 40 pp. Available at: http://extranet.who.int/iris/bitstream/123456789/669/1/9789241596657_eng.pdf.

de Bon, H., Parrot, L. and Moustier, P. 2010. Sustainable urban agriculture in developing countries. A review. *Agronomy for Sustainable Development*, 30 (1): 21–32. Available at: http://dx.doi.org/10.1051/agro:2008062.

de Fraiture, C., Giordano, M. and Liao, Y. 2008. Biofuels and implications for agricultural water use: blue impacts of green energy. *Water Policy*, 10 (S1): 67–81. Available at: http://www.iwmi.cgiar.org/EWMA/files/papers/Biofuels-Charlotte.pdf.

de Schutter, O. 2008. Building resilience: a human rights framework for world food and nutrition security. New York: United Nations, Human Rights Council, 43 pp. Available at: http://www.srfood.org/images/stories/pdf/officialreports/or1-a-1-hrc-9-23final-eng.pdf.

de Schutter, O. 2010. Report submitted by the Special Rapporteur on the right to food, Olivier De Schutter on 20 December 2010. New York: United Nations. (GE.1017850 (F) 280111 020211 – A/HRC/16/49), 23 pp. Available at: http://www.srfood.org/images/stories/pdf/officialreports/20110308_a-hrc-16-49_agroecology_fr.pdf.

de Schutter, O. 2011. G20: cinq priorités pour améliorer la sécurité alimentaire mondiale. *Le Monde*, 8 June.

Delisle, H. 2010. Findings on dietary patterns in different groups of African origin undergoing nutrition transition. *Applied Physiology, Nutrition, and Metabolism*, 35 (2): 224–228. Available at: http://dx.doi.org/10.1139/H10-008.

Delormier, T., Frohlich, K. L. and Potvin, L. 2009. Food and eating as social practice understanding eating patterns as social phenomena and implications for public health. *Sociology of Health and Illness*, 31 (2): 215–228. Available at: http://dx.doi.org/10.1111/j.1467-9566.2008.01128.x.

Demeke, M., Pangrazio, G. and Maetz, M. 2009. *Country Responses to the Food Security Crisis: Nature and Preliminary Implications of the Policies Pursued*. Rome: FAO. 29 pp. Available at: http://www.globalbioenergy.org/uploads/media/0812_FAO_-_Country_responses_to_the_crisis.pdf.

Desclaux, D., Chiffoleau, Y. and Nolot, J. M. 2009. Pluralité des agricultures biologiques : enjeux pour la construction des marchés, le choix des variétés et les schémas d'amélioration des plantes. Diversity of organic farming systems: challenges for the construction of markets, varieties and plant breeding techniques. *Innovations Agronomiques*, 4: 297–306. Available at: http://www.inra.fr/ciag/revue_innovations_agronomiques/volume_4_janvier_2009.

Deverre, C. and Lamine, C. 2010. Les systèmes agroalimentaires alternatifs. Une revue de travaux anglophones en sciences sociales. *Economie rurale* (317): 57–73. Available at: http://economierurale.revues.org/index2676.html#quotation.

Dhérissard, G. and Viel, D. 2007. À problème complexe, solutions complexes. *Economie et Humanisme* (380): 9–12. Available at: http://www.revue-economie-et-humanisme.eu/bdf/docs/r380_9_problemesolutionscomplexes.pdf.

Diasio, N., Hubert, A. and Pardo, V. 2009. Alimentations adolescentes en France: principaux résultats du programme de recherche AlimAdos. Paris: OCHA (Cahiers de l'Ocha), 220 pp. Available at: http://www.lemangeur-ocha.com/catalogue/auteur-ouvrage/0/alimentations-adolescentes-en-france/disp/cahiers-de-locha/.

Doak, C., Adair, L., Monteiro, C. and Popkin, B. 2000. Overweight and underweight coexist within households in Brazil, China and Russia. *Journal of Nutrition*, 130 (2): 2965–2971. Available at: http://jn.nutrition.org/content/130/12/2965.full.pdf.

Doak, C. M., Adair, L. S., Bentley, M., Monteiro, C. and Popkin, B. M. 2004. The dual burden household and the nutrition transition paradox. *International Journal of Obesity*, 29 (1): 129–136. Available at: http://dx.doi.org/10.1038/sj.ijo.0802824.

Donald, B., Gertler, M., Gray, M. and Lobao, L. 2010. Re-regionalizing the food system? *Cambridge Journal of Regions, Economy and Society*, 3 (2): 171–175. Available at: http://cjres.oxfordjournals.org/cgi/content/short/rsq020v1.

Dorin, B. and Gitz, V. 2008. Ecobilans de biocarburants: une revue des controverses. *Natures Sciences Societes*, 16 (4): 337–347. Available at: http://dx.doi.org/10.1051/nss/2008064.

Drewnowski, A. and Popkin, B. M. 1997. The nutrition transition: new trends in the global diet. *Nutrition Reviews*, 55 (2): 31–43. Available at: http://dx.doi.org/10.1111/j.1753-4887.1997.tb01593.x.

Dubuisson-Quellier, C. 2008. Consumer involvement in fair trade and local food systems: delegation and empowerment regimes. *GeoJournal*, 73 (1): 55–65. Available at: http://dx.doi.org/10.1007/s10708-008-9178-0.

Dubuisson-Quellier, C. 2009a. La consommation comme pratique sociale. In P. Steiner and F. Vatin, eds, *Traité de Sociologie Économique*. Paris: PUF, pp. 727–776.

Dubuisson-Quellier, C. 2009b. *La consommation engagée*. Paris: Les presses de Sciences Po (*Collection Contester*), 143 pp. Available at: http://books.google.fr/books?id=42hWPgAACAAJ.

Dubuisson-Quellier, S. 2010. From consumerism to the empowerment of consumers: the case of consumer oriented movements in France. *Sustainability*, 2 (7): 1849–1868. Available at: http://www.mdpi.com/2071-1050/2/7/1849/pdf.

Duchin, F. 2005. Sustainable consumption of food: a framework for analyzing scenarios about changes in diets. *Journal of Industrial Ecology*, 9 (1–2): 99–114. Available at: http://dx.doi.org/10.1162/1088198054084707.

Dury, S., Medou, J. C., Foudjem-Tita, D. and Nolte, C. 2004. Limites du système local d'approvisionnement alimentaire urbain en Afrique subsaharienne: le cas des amylacées au Sud-Cameroun. *Cahiers Agricultures*, 13 (1): 116–124.

Available at: http://www.john-libbey-eurotext.fr/fr/revues/medecine/hma/e-docs/00/03/FE/EB.

Earles, J. and Halog, A. 2011. Consequential life cycle assessment: a review. *The International Journal of Life Cycle Assessment*, 16 (5): 445–453. Available at: http://dx.doi.org/10.1007/s11367-011-0275-9.

European Environment Agency 2007. *Greenhouse Gas Emission Trends and Projections in Europe 2007*. Copenhagen: European Environment Agency (EEA Report n° 5/2007), 102 pp. Available at: http://www.eea.europa.eu/publications/eea_report_2007_5/Greenhouse_gas_emission_trends_and_projections_in_Europe_2007.pdf.

EFSA 2010. Scientific opinion on establishing food-based dietary guidelines. *EFSA Journal*, 8 (3): 42 pp. Available at: http://www.efsa.europa.eu/fr/efsajournal/doc/1460.pdf.

Elghali, L., Clift, R., Begg, K. G. and McLaren, S. 2008. Decision support methodology for complex contexts. *Proceedings of the Institution of Civil Engineers-Engineering Sustainability*, 161 (1): 7–22. Available at: http://dx.doi.org/10.1680/ensu.2008.161.1.7.

Elinder, L. S. 2005. Obesity, hunger, and agriculture: the damaging role of subsidies. *BMJ*, 331 (7528): 1333–1336. Available at: http://dx.doi.org/10.1136/bmj.331.7528.1333.

Environmental Protection Agency 2011. *Inventory of U.S. Greenhouse Gas Emissions and Sinks: 1990–2009*. Washington: Environmental Protection Agency (US Greenhouse Gas Inventory Report no. 430-R-11–005), 459 pp. Available at: http://www.epa.gov/climatechange/Downloads/ghgemissions/US-GHG-Inventory-2011-Complete_Report.pdf.

Ertug, F. 2006. *Proceedings of the IVth International Congress of Ethnobotany ICEB 2005. Yeditepe University, Istanbul, 21–26 August 2005*. Istanbul, Turkey: Zero Prod. Ltd., 536 pp. Available at: http://www.tropicallab.ugent.be/melissa.pdf.

Escalon, H., Bossard, C., Beck, F. and Bachelot-Narquin, R. P. 2009. *Baromètre Nutrition Santé: Edition 2008*. Paris: INPES, 419 pp. Available at: http://www.inpes.sante.fr/CFESBases/catalogue/pdf/1270.pdf.

Estaquio, C., Castetbon, K., Kesse-Guyot, E. *et al.* 2008. The French National Nutrition and Health Program score is associated with nutritional status and risk of major chronic diseases. *Journal of Nutrition*, 138 (5): 946–953. Available at: http://jn.nutrition.org/content/138/5/946.full.

Etievant, P., Bellisle, F., Dallongeville, J. *et al.* (eds), 2010. Dietary behaviours and practices: determinants, action, outcomes. Executive summary of the Collective Scientific Expertise conducted by INRA, France, 62 pp. Available at: http://www.international.inra.fr/the_institute/scientific_expertise/expert_reports/dietary_behaviours.

European Commission 2005. *Impact Assessment Guidelines*. Brussels: European Commission.

European Environment Agency 2006. How much bioenergy can Europe produce without harming the environment? EEA report No 7/2006, 72 pp.

Evans, A. 2009. The feeding of the nine billion: global food security for the 21st Century. Chatham House Report, Chatham House, 61 pp. Available at: http://www.chathamhouse.org.uk/publications/papers/download/-/id/694/file/13179_r0109food.pdf.

Evans, A. 2010. *Gloabilization and Scarcity: Multilateralism for a World with Limits*. New York: Center on International Cooperation. 62 pp. Available at: http://reliefweb. int/sites/reliefweb.int/files/resources/0B5571994F2850CB492577DE0024 A12D-Full_Report.pdf.

Fan, X., Bourg, D. and Erkman, S. 2006. L'économie circulaire en Chine. Vers une prise en compte de l'environnement dans le système économique chinois? *Futuribles* (324): 21–41. Available at: http://dx.doi.org/10.1051/futur:200632421.

FAO 1981. Food loss prevention in perishable crops. Agricultural Service Bulletin, no. 43. Rome: FAO, 72 pp. Available at: http://www.fao.org/docrep/s8620e/S8620E00. htm#Contents.

FAO 1996. Rapport du Sommet mondial de l'alimentation. Rome: FAO, 261 pp. Available at: http://www.fao.org/docrep/003/w3548f/w3548f00.htm.

FAO 1998. The state of food and agriculture 1998, FAO Agriculture Series, no. 31. Rome: FAO, 371 pp. Available at: http://www.fao.org/docrep/w9500e/w9500e00.htm.

FAO 2001. Human energy requirements. Report of a Joint FAO/WHO/UNU Expert Consultation. Food and Nutrition Technical Report Series. Rome: FAO, 103 pp. Available at: ftp.fao.org/docrep/fao/007/y5686e/y5686e00.pdf.

FAO 2006. L'état de l'insécurité alimentaire dans le monde 2006. Rome: FAO, 44 pp. Available at: ftp://ftp.fao.org/docrep/fao/009/a0750f/a0750f.pdf.

FAO 2007. Post-harvest loss compendium: database on post-harvest losses of different products. Rome: FAO. Available at: http://www.fao.org/inpho.

FAO 2008a. La situation mondiale de l'alimentation et de l'agriculture. Les biocarburants: perspectives, risques et opportunité. Rome: FAO, 156 pp. Available at: ftp://ftp.fao.org/docrep/fao/011/i0100f/i0100f.pdf.

FAO 2008b. Les silos métalliques familiaux: alliés essentiels de la FAO dans la lutte contre la faim. Rome: FAO, 8 pp. Available at: ftp://ftp.fao.org/docrep/fao/011/ i0100f/i0100f.pdf.

FAO 2009a. La situation mondiale de l'alimentation et de l'agriculture: Le point sur l'élevage. Rome: FAO, 202 pp. Available at: http://www.fao.org/docrep/012/i0680f/ i0680f.pdf.

FAO 2009b. More people than ever are victims of hunger. Rome: FAO. Press release, 4 pp. Available at: http://www.fao.org/fileadmin/user_upload/newsroom/docs/ Press release june-en.pdf.

FAO 2010a. *Bioenergy and Food Security: The BEFS Analytical Framework. Environment and Natural Resources Management Series*. Rome: Bioenergy and Food Security Project, Food and Agriculture Organization of the United Nations. Available at: http:// www.fao.org/docrep/013/i1968e/i1968e.pdf.

FAO 2010b. Definition of sustainable diets. International Scientific Symposium on Biodiversity and Sustainable Diets. United Against Hunger. Rome: FAO, 2 pp. Available at: http://www.fao.org/ag/humannutrition/23781-0e8d8dc364ee 46865d5841c48976e9980.pdf.

FAO 2010c. FAO hunger statistics. Available at: http://reliefweb.int/sites/reliefweb.int/ files/resources/5506BABD1F27BFC0C12577490047D159-FAO_Mar2010.pdf, accessed 11 July 2011.

FAO 2010d. Food for the cities: aliments pour les villes. Available at: http://www.fao.
org/fcit/fr/.

FAO and World Bank 2010. Reducing post-harvest losses in grain supply chains in
Africa: lessons learned and practical guidelines. Rome and Washington DC:
FAO and World Bank, 120 pp. Available at: http://reliefweb.int/sites/
reliefweb.int/files/resources/5506BABD1F27BFC0C12577490047D159-
FAO_Mar2010.pdf.

FAO-UNO 2009. The state of food insecurity in the world: economic crises: impacts
and lessons learned. Rome: FAO, 61 pp. Available at: ftp://ftp.fao.org/docrep/fao/
012/i0876e/i0876e.pdf.

Favier, J., Ireland-Ripert, J., Toque, C. and Feinberg, M. 1995. CIQUAL Répertoire géné
ral des aliments. Table de composition. [General repertoire of foods. Food
composition table.] 2nd edn. Paris: Lavoisier, Tec&Doc.

Faye, B. 2011. Quelles espèces pour quels espaces ? Quels systèmes pour quels
territoires ? Les enjeux de la durabilité en élevage laitier. *Actes du colloque de
l'Ocha Cultures des Laits du Monde, 6–7 May 2010. Paris*. Paris: OCHA, pp. 160–173.
Available at: http://www.lemangeur-ocha.com/fileadmin/images/dossiers/
Laits-du-monde-S3-BF.pdf.

Faye, B. and Duteurtre, G. 2009. *L'élevage, richesse des pauvres*. Versailles, France:
Editions Quae, 288 pp.

Feagan, R. 2007. The place of food: mapping out the 'local' in local food systems.
Progress in Human Geography, 31 (1): 23–42. Available at: http://phg.sagepub.com/
cgi/content/abstract/31/1/23.

Feeney, E., O'Brien, S., Scannell, A., Markey, A. and Gibney, E. R. 2011. Genetic
variation in taste perception: does it have a role in healthy eating? *Proceedings of
the Nutrition Society*, 70 (01): 135–143. Available at: http://dx.doi.org/10.1017/
S0029665110003976.

Feillet, P. 2007. *La nourriture des Français. De la maîtrise du feu ... aux années 2030*.
Versailles, France: Editions Quae, 245 pp.

Figuié, M. and Bricas, N. 2008. Equité internationale, la surresponsabilisation des
consommateurs. *Courrier de la planète* (87): 41. Available at: http://publications.
cirad.fr/une_notice.php?dk=549690.

Figuié, M. and Fournier, T. 2010. Risques sanitaires globaux et politiques nationales:
la gestion de la grippe aviaire au Vietnam. *Revue d'Etudes en Agriculture et
Environnement*, 91 (3): 327–343. Available at: http://www.raestud.eu/pdf/REAE-93-
1-Figuie-Fournier.pdf.

FINE 2001. Fairtrade International code of practice. Available at: http://www.fairtrade.
net/about_fairtrade.html, accessed April 2008.

Finucane, M. M., Stevens, G. A., Cowan, M. J. *et al*. 2011. National, regional, and global
trends in body-mass index since 1980: systematic analysis of health examination
surveys and epidemiological studies with 960 country-years and 9.1 million
participants. *The Lancet*, 377 (9765): 557–567. Available at: http://dx.doi.org/
10.1016/S0140-6736(10)62037-5.

Fischler, C. 1990. *L'homnivore: le goût, la cuisine et le corps*. Paris: Odile Jacob, 414 pp.

Fischler, C. and Masson, E. 2008. *Manger: Français, Européens et Américains face à l'alimentation*. Paris: Odile Jacob, 336 pp.

Fischer, G., Hizsnyik, E., Prieler, S., Shah, M. and van Velthuizen, H. 2009. *Biofuels and Biodiversity. International Institute for Applied Systems Analysis (IIASA)*. Luxembourg: International Institute for Applied Systems Analysis (IIASA), 223 pp.

Flatt, J. P. 2011. Issues and misconceptions about obesity. *Obesity*, 19 (4): 676–686. Available at: http://dx.doi.org/10.1038/oby.2011.7.

Flores, M., Macias, N., Rivera, M. *et al*. 2010. Dietary patterns in Mexican adults are associated with risk of being overweight or obese. *Journal of Nutrition*, 140 (10): 1869–1873. Available at: http://dx.doi.org/110.3945/jn.110.121533.

Fogel, R. W. 1994. Economic growth, population theory, and physiology: the bearing of long-term processes on the making of economic policy. *American Economic Review*, 84 (3): 369–395. Available at: http://www.nber.org/papers/w4638.pdf.

Foresight 2011. The future of food and farming: challenges and choices for global sustainability. Final Project. London: The Government Office for Science, 208 pp. Available at: http://www.bis.gov.uk/assets/bispartners/foresight/docs/food-and-farming/11-546-future-of-food-and-farming-report.

Fouilleux, E. 2010. Les standards volontaires, instruments montants des politiques agricoles et alimentaires. Entre internationalisation et privatisation In B. Hervieu, ed. *Les Mondes Agricoles en Politique*. Paris: Presses de Science Po, pp. 371–396.

Fransen, L. W. and Kolk, A. 2007. Global rule-setting for business: a critical analysis of multi-stakeholder standards. *Organization*, 14 (5): 667–684. Available at: http://org.sagepub.com/content/14/5/667.abstract.

Freibauer, A., Mathijs, E., Brunori, G. *et al*. 2011. *The 3rd SCAR Foresight Exercise: Sustainable Food Consumption and Production in a Resource-constrained World*. Brussels: European Commission–Standing Committee on Agricultural Research (SCAR), 149 pp. Available at: http://ec.europa.eu/research/agriculture/scar/pdf/scar_feg3_final_report_01_02_2011.pdf.

Friedmann, H. 2009. Discussion: moving food regimes forward: reflections on symposium essays. *Agriculture and Human Values*, 26 (4): 335–344. Available at: http://dx.doi.org/10.1007/s10460-009-9225-6.

Fritz, M. and Schiefer, G. 2008. Food chain management for sustainable food system development: a European research agenda. *Agribusiness*, 24 (4): 440–452. Available at: http://www3.interscience.wiley.com/journal/121460878/abstract.

Fujita, M. and Thisse, J.-F. 2002. *Economics of Agglomeration: Cities Industrial Location, and Regional Growth*. Cambridge, UK and New York: Cambridge University Press, 466 pp.

Fulponi, L. 2006. Private voluntary standards in the food system: The perspective of major food retailers in OECD countries. *Food Policy*, 31 (1): 1–13. Available at: http://www.sciencedirect.com/science/article/pii/S030691920500045X.

Gaigné, C., Riou, S. and Thisse, J.-F. 2011a. Are compact cities environmentally friendly? CEPR Discussion Papers. Available at: http://www.cepr.org/pubs/dps/DP8297.asp.

Gaigné C., Le Gallo J., Larue S. and Schmitt, B. 2011b. Does regulation of manure land application work against agglomeration economies? Theory and evidence from the

French hog sector. *American Journal of Agricultural Economics*, 94 (1): 116–132. Available at: http://dx.doi.org/10.1093/ajae/aar121.

Garnett, T. 2011. Where are the best opportunities for reducing greenhouse gas emissions in the food system (including the food chain)? *Food Policy*, 36 (Supplement 1): S23–S32. Available at: http://dx.doi.org/10.1016/j.foodpol.2010.10.010.

Garrett, J. and Ruel, M. 2003. Stunted child–overweight mother pairs: an emerging policy concern? *Food Consumption and Nutrition Division*, Discussion Paper No. 148: 36 pp. Available at: http://citeseerx.ist.psu.edu/viewdoc/summary?doi=10.1.1.58.3600&rank=1.

Gauvrit, L. and Mora, O. 2010. *Prospective Vega: les usages non alimentaires de la biomasse végétale à l'horizon 2050*. Paris: INRA-DEPE, 91 pp.

Gérard, F., Alpha, A., Beaujeu, R. *et al.* 2011. *Managing Food Price Volatility for Food Security and Development*. Paris: Groupe de Recherches et d'Echanges sur la Régulation des Marchés Agricoles (GREMA), 161 pp. Available at: http://www.inter-reseaux.org/IMG/pdf/Managing_Food_Price_Volatility_for_Food_Security_and_Development_Grema.pdf.

Getz, C. and Shreck, A. 2006. What organic and Fair Trade labels do not tell us: towards a place-based understanding of certification. *International Journal of Consumer Studies*, 30 (5): 490–501. Available at: http://www.blackwell-synergy.com/doi/pdf/10.1111/j.1470-6431.2006.00533.x.

Gilbert, C. 2007. *Les Crises Sanitaires de Grande Ampleur: Un Nouveau Défi?* Paris: La Documentation Française, 70 pp.

Gilbert, C. L. and Morgan, C. W. 2010. Food price volatility. *Philosophical Transactions of the Royal Society B: Biological Sciences*, 365 (1554): 3023–3034. Available at: http://rstb.royalsocietypublishing.org/content/365/1554/3023.abstract.

Giovannucci, D., Varangis, P. and Larson, D. 2000. Warehouse receipts: facilitating credit and commodity markets. Washington DC: The World Bank, 8 pp. Available at: http://www-wds.worldbank.org/external/default/WDSContentServer/WDSP/IB/2007/07/10/000090341_20070710160223/Rendered/PDF/401220Warehouse1Receipts01PUBLIC1.pdf.

Gluckman, P. D., Hanson, M. A., Bateson, P. *et al.* 2009. Towards a new developmental synthesis: adaptive developmental plasticity and human disease. *The Lancet*, 373 (9675): 1654–1657. Available at: http://dx.doi.org/10.1016/S0140-6736(09)60234-8.

Godard, O. 2009. Les enjeux du développement durable. duALIne oral communication.

Gojard, S. 2000. L'alimentation dans la prime enfance. Diffusion et réception des normes de puériculture. *Revue Francaise De Sociologie*, 3: 475–512. Available at: http://www.persee.fr/articleAsPDF/rfsoc_0035-2969_2000_num_41_3_5291/article_rfsoc_0035-2969_2000_num_41_3_5291.pdf.

Gonzalez, C. A. and Riboli, E. 2010. Diet and cancer prevention: contributions from the European prospective investigation into cancer and nutrition (EPIC) study. *European Journal of Cancer*, 46 (14): 2555–2562. Available at: http://dx.doi.org/10.1016/j.ejca.2010.07.025.

Goodman, D. 1997. World-scale processes and agro-food systems: critique and research needs. *Review of International Political Economy*, 4 (4): 663–687. Available at: http://www.informaworld.com/10.1080/09672299708565787.

Gopal, V., Prabakaran, S. and Balasubramanian, P. R. 1996. Effect of a biogas-plant effluent-based pelleted diet on the growth of *Oreochromis mossambicus* fingerlings. *Bioresource Technology*, 58 (3): 315–317. Available at: http://dx.doi.org/10.1016/s0960-8524(96)00116-2.

Gordon, T. 2005. Fair trade organic products conquer European markets. Fair trade and organic standards. *Entwicklung + Landlicher Raum*, 39 (3): 24–26.

Gorgens, T., Meng, X. and Vaithianathan, R. 2011. Stunting and selection effects of famine: a case study of the Great Chinese Famine. *Journal of Development Economics*, 97 (1): 99–111. Available at: http://dx.doi.org/10.1016/j.jdeveco.2010.12.005.

Gouel, C. and Jean, S. 2011. Optimal food price stabilisation in a small open developing country. New York: World Bank, 33 pp. Available at: http://pagesperso.dial.prd.fr/dial_pagesperso/dial_eve_shocks/pdf/185_S%C3%A9bastien Jean.pdf.

Goulet, J., Lamarche, B., Nadeau, G. and Lemieux, S. 2003. Effect of a nutritional intervention promoting the Mediterranean food pattern on plasma lipids, lipoproteins and body weight in healthy French-Canadian women. *Atherosclerosis*, 170 (1): 115–124. Available at: http://dx.doi.org/10.1016/S0021-9150(03)00243-0.

Granatstein, D. and Kupferman, E. 2008. Sustainable horticulture in fruit production. *International Society for Horticultural Science (ISHS)*, 295–308. Available at: http://www.actahort.org/.

Gray, S. 2009. Down the drain. Quantification and exploration of food and drink waste disposed of to the sewer by households in the UK. Working paper, WRAP, 56 pp. Available at: http://www.wrap.org.uk/downloads/Down_the_drain_-_report.e1b8e60a.8049.pdf.

Griffon, M. E. 2006. *Nourrir la planète: Pour une révolution doublement verte*. Paris: Odile Jacob, 456 pp.

Grimm, E. R. and Steinle, N. I. 2011. Genetics of eating behavior: established and emerging concepts. *Nutrition Reviews*, 69 (1): 52–60. Available at: http://dx.doi.org/10.1111/j.1753-4887.2010.00361.x.

Grolleau, G. and Caswell, J. A. 2006. Interaction between food attributes in markets: the case of environmental labeling. *Journal of Agricultural and Resource Economics*, 31 (3): 471–484. Available at: http://ageconsearch.umn.edu/handle/8636.

Grolleaud, M. 2002. Pertes après récolte: un concept mal défini ou mal utilisé. Étude synthétique et didactique sur le phénomène des pertes se produisant tout au long du système après récolte. Rome: FAO, n.p. Available at: http://www.fao.org/docrep/004/AC301F/AC301f00.HTM.

Guignon, N., Collet, M., Gonzales, L. *et al.* 2010. La santé des enfants en grande section de maternelle en 2005–2006. *Etudes & Résultats* (737): 1–8. Available at: http://www.sante-sports.gouv.fr/IMG/pdf/er737-2.pdf.

Günther, D. and Zimprich, E. 1997. Gender-orientation in the post-harvest sector. Pointers for identifying gender-specific aspects in the post-harvest sector. Bonn,

Germany: Federal Ministry for Economic Cooperation and Development (BMZ), 35 pp. Available at: http://www.cd3wd.com/cd3wd_40/INPHO/VLIBRARY/ GTZHTML/X0285E/EN/X0285E00.HTM.

Gustavsson, J., Cederberg, J., Sonesson, J., van Otterdijk, J. and Meybeck, A. 2011. *Global Food Losses and Food Waste: Extent, Causes and Prevention.* Rome: FAO, 29 pp. Available at: http://www.fao.org/fileadmin/user_upload/ags/publications/GFL_web.pdf.

Haberl, H., Erb, K.-H., Krausmann, F. *et al.* 2011. Global bioenergy potentials from agricultural land in 2050: Sensitivity to climate change, diets and yields. *Biomass and Bioenergy*, 35 (12): 4753–4769. Available at: http://dx.doi.org/10.1016/j. biombioe.2011.04.035.

Halkier, B. 2009. A practice theoretical perspective on everyday dealings with environmental challenges of food consumption. *Anthropology of Food* (5). Available at: http://aof.revues.org/index6405.html.

Halweil, B. 2006. L'agriculture biologique peut-elle nous nourrir tous? *L'Etat de la planète* (27): 9 pp. Available at: http://www.delaplanete.org/ L-agriculture-biologique-peut-elle.html.

Hardesty, S. D. 2008. The growing role of local food markets. *American Journal of Agricultural Economics*, 90 (5): 1289–1295. Available at: http://ajae.oxfordjournals. org/content/90/5/1289.short.

Havaligi, N. 2009. Climate change and food security in cities. *5th Urban Research Symposium 2009*, 18 pp. Available at: http://siteresources.worldbank.org/ INTURBANDEVELOPMENT/Resources/336387-1256566800920/6505269- 1268260567624/Havaligi.pdf.

Hayami, Y. and Ruttan, V. W. 1985. *Agricultural Development: An International Perspective.* Baltimore, MD: Johns Hopkins University Press, 506 pp.

Hayes, J. E., Sullivan, B. S. and Duffy, V. B. 2010. Explaining variability in sodium intake through oral sensory phenotype, salt sensation and liking. *Physiology & Behavior*, 100 (4): 369–380. Available at: http://siteresources. worldbank.org/INTURBANDEVELOPMENT/Resources/336387-1256566800920/ 6505269-1268260567624/Havaligi.pdf.

Headey, D. 2011. Was the global food crisis really a crisis? Simulations versus self-reporting. Washington DC: IFPRI, 72 pp. Available at: http://siteresources. worldbank.org/INTURBANDEVELOPMENT/Resources/336387-1256566800920/ 6505269-1268260567624/Havaligi.pdf.

Heinberg, R. and Bomford, M. 2009. *The Food and Farming Transition: Toward a Post Carbon Food System.* Sebastopol, CA: Post Carbon Institute. 41 pp. Available at: http://www.postcarbon.org/reports/PCI-F&FTransition-highrez.pdf.

Heller, M. C. and Keoleian, G. A. 2000. Life cycle-based sustainability indicators for assessment of the US food system. Report, Center for Sustainable Systems, School of Natural Resources and Environment, University of Michigan, 59 pp. Available at: http://www.redwoodcity.org/manager/initiatives/gardens/more_reading/Life Cycle Based Sustainability Indicators for Assessment of the US Food System.pdf.

Henkes, C. 1992. Investigations into insect population dynamics, damage and losses of stored maize: an approach to IPM in small farms in Tanzania with special reference to *Prostephanus truncatus* (Horn). PhD thesis. Technische Universitaet Berlin, Berlin, 124 pp.

Henneberry, S. R., Whitacre, B. and Agustini, H. N. 2009. An evaluation of the economic impacts of Oklahoma farmers markets. *Journal of Food Distribution Research*, 40 (3): 64–78. Available at: http://purl.umn.edu/99760.

Herrero, M., Thornton, P. K., Notenbaert, A. M. *et al.* 2010. Smart investments in sustainable food production: revisiting mixed crop-livestock systems. *Science*, 327 (5967): 822–825. Available at: http://www.sciencemag.org/content/327/5967/822.short.

Herrmann, H. 1991. Seguridad alimentaria: Comparación de impactos socio-económicos en la tenencia del silo metálico versus sistema tradicional de almacenamiento. Research report (Project Document 09.03. PROJ014 DOC91.02), 80 pp. + annexes. Available at: http://www.postcosecha.net/en/Home/document.php?itemID=4398&langID=3.

Herrmann, R. and Röder, C. 1995. Does food consumption converge internationally? Measurement, empirical tests and determinants. *European Review of Agricultural Economics*, 22 (3): 400–414. Available at: http://erae.oxfordjournals.org/content/22/3/400.abstract.

Hill, J. O. 2009. Can a small-changes approach help address the obesity epidemic? A report of the Joint Task Force of the American Society for Nutrition, Institute of Food Technologists, and International Food Information Council. *American Journal of Clinical Nutrition*, 89 (2): 477–484. Available at: http://erae.oxfordjournals.org/content/22/3/400.abstract.

Hoddinott, J., Maluccio, J. A., Behrman, J. R., Flores, R. and Martorell, R. 2008. Effect of a nutrition intervention during early childhood on economic productivity in Guatemalan adults. *The Lancet*, 371 (9610): 411–416. Available at: http://www.sciencedirect.com/science/article/B6T1B-4RR1YFB-14/2/e931676fcd2ccacaec6abcedc4f9f560.

Hoekstra, A. Y. 2002. Virtual water trade: proceedings of the international expert meeting on virtual water trade. *International Expert Meeting on Virtual Water Trade*. Delft, The Netherlands, 12–13 December 2002, 242 pp. Available at: http://www.waterfootprint.org/Reports/Report12.pdf.

Holt, G. and Amilien, V. 2007. Special issue on local food products and systems. *Anthropology of Food*, 6: 405–539. Available at: http://aof.revues.org/sommaire402.html.

Hubert, B. 2002. Sustainable development: think forward and act now. Agricultures and sustainable development. The stakes of knowledge and research attitudes. In P. Legrand, A. Fraval and C. Laurent, eds, *Johannesburg. INRA Faced with Sustainable Development: Landmarks for the Johannesburg Conference*, Dossiers de l'Environnement de l'INRA, no. 22. Paris: INRA Editions, pp. 41–56.

Hubert, J. P. 2009. Dans les grandes agglomérations, la mobilité quotidienne des habitants diminue, et elle augmente ailleurs. INSEE Première, no.1252, 4 pp. Available at: http://www.insee.fr/fr/ffc/ipweb/ip1252/ip1252.pdf.

Hugon, P. 1985. Le miroir sans tain. Dépendance alimentaire et urbanisation en Afrique: un essai d'analyse en termes de filières. In ALTERSIAL, ENSIA-GRET, CERED,

CERNEA, MSA, ORSTOM, N. Bricas, eds, *Nourrir les villes en Afrique sub-Saharienne*. Paris: L'Harmattan, pp. 9–46. Available at: http://www.sudoc.abes.fr/DB=2.1/SET=4/TTL=1/CLK?IKT=1016&TRM=Nourrir+les+villes+en+Afrique+sub-saharienne.

Huijbregts, P., Feskens, E., Rasanen, L. *et al.* 1997. Dietary pattern and 20 year mortality in elderly men in Finland, Italy, and the Netherlands: longitudinal cohort study. *British Medical Journal*, 315 (7099): 13–17. Available at: http://pubmedcentralcanada.ca/picrender.cgi?accid=PMC2127011&blobtype=pdf.

IEA 2008. *Energy Technology Perspectives 2010: Scenarios & Strategies to 2050*. Paris: International Energy Agency, 643 pp. Available at: http://www.iea.org/textbase/nppdf/free/2008/etp2008.pdf.

Ilbery, B., Morris, C., Buller, H., Maye, D. and Kneafsey, M. 2005. Product, process and place: an examination of food marketing and labelling schemes in Europe and North America. *European Urban and Regional Studies*, 12 (2): 116–132. Available at: http://dx.doi.org/10.1177/0969776405048499.

IMF 2011. *World Economic Outlook*. Washington DC: International Monetary Fund, 221 pp. Available at: http://www.scribd.com/doc/53986940/IMF-World-Economic-Outlook-April-2011.

INRA 2010. Document d'orientation INRA 2010–2020: une science pour l'impact. Paris: INRA, 57 pp. Available at: http://www.inra.fr/l_institut/missions_et_strategie/la_strategie/document_d_orientation_2010_2020.

INSEE 2006. La consommation des ménages en 2006. Paris: INSEE. Available at: http://www.insee.fr/fr/publications-et-services/irweb.asp?id=conso2006.

Institut Français du Pétrole (IFP) 2012. Quel avenir pour le pétrole (What future for oil?) at: http://www.ifp.fr.

Issa, C., Darmon, N., Salameh, P. *et al.* 2011. A Mediterranean diet pattern with low consumption of liquid sweets and refined cereals is negatively associated with adiposity in adults from rural Lebanon. *International Journal of Obesity*, 35 (2): 251–258. Available at: http://dx.doi.org/10.1038/ijo.2010.130.

Jacquet, F., Bamiere, L., Bureau, J. C. *et al.* 2007. Les enjeux du développement des biocarburants dans l'Union européenne. *INRA Sciences Sociales Recherches en Economie et Sociologie Rurales* 2–3: 1–6. Available at: http://www.inra.fr/internet/Departements/ESR/publications/iss/pdf/iss07-2Jac.pdf?PHPSESSID=9a464bd90ff3f4bda8e7c214d9d92f8f.

Jacquet, J., Pauly, D., Ainley, D. *et al.* 2010. Seafood stewardship in crisis. *Nature*, 467 (7311): 28–29. Available at: http://dx.doi.org/10.1038/467028a.

Jacquiau, C. 2006. Les coulisses du commerce équitable: mensonges et vérités sur un petit business qui monte. Paris: Mille et Une Nuits, 484 pp.

Jayne, T. S. and Tschirley, D. L. 2009. Food price spikes and strategic interactions between the public and private sectors: market failures or governance failures? *Commodity Market Review 2009–2010*. Rome: FAO, pp. 3–18. Available at: http://ageconsearch.umn.edu/bitstream/97142/2/Chapter_Jayne.pdf.

Jeanneret, P., Baumgartner, D. U., Freiermuth Knuchel, R. and Gaillard, G. 2008. Integration of biodiversity as impact category for LCA in agriculture (SALCA-Biodiversity). 6th International Conference on LCA in the Agri-Food

Sector Zurich, 12–14 November, 2008, 6 pp. Available at: http://www.agroscope. admin.ch/data/publikationen/1296463076_Jeanneret_et_al_Agri_Food.pdf.

Jeffries, P. and Jeger, M. J. 1990. The biological control of post-harvest diseases of fruit. *Post-harvest News Info*, 1: 365–368.

Jeon, Y. W. and Halos, L. S. 1991. Addressing R&D for cassava postharvest system in West Africa. ASAEN Meeting Presentation, 9 pp.

Jones, T. W. 2004. Using contemporary archaeology and applied anthropology to understand food loss in the American food system. Tucson: University of Arizona, Bureau of Applied Research in Anthropology, 6 pp. Available at: http://www.ce.cmu.edu/gdrg/readings/2006/12/19/Jones_UsingContemporary ArchaeologyAndAppliedAnthropologyToUnderstandFoodLossInAmerican FoodSystem.pdf.

Jørgensen, A., Hauschild, M. Z., Jorgensen, M. S. and Wangel, A. 2009. Relevance and feasibility of social life cycle assessment from a company perspective. *International Journal of Life Cycle Assessment*, 14 (3): 204–214. Available at: http://dx. doi.org/10.1007/s11367-009-0073-9.

Kadim, O. S. 2009. Le stockage des produits végétaux au Maroc: choix technique et pertes post-récolte. PhD thesis (Sciences de la vie et de l'environnement). Université de Franche-comté, Besançon, 81 pp.

Kant, A. K. 1996. Indexes of overall diet quality: a review. *Journal of the American Dietetic Association*, 96 (8): 785–791. Available at: http://dx.doi.org/10.1016/ S0002-8223(96)00217-9.

Kanter, C., Messer, K. D. and Kaiser, H. M. 2009. Does production labeling stigmatize conventional milk? *American Journal of Agricultural Economics*, 91 (4): 1097–1109. Available at: http://ajae.oxfordjournals.org/content/91/4/1097. abstract.

Keesing, F., Belden, L. K., Daszak, P. *et al.* 2010. Impacts of biodiversity on the emergence and transmission of infectious diseases. *Nature*, 468 (7324): 647–652. Available at: http://dx.doi.org/10.1038/nature09575.

Kelleher, K. 2005. Discards in the world's marine fisheries. FAO Fisheries Technical Paper. Rome: FAO, 131 pp. Available at: ftp://ftp.fao.org/docrep/fao/008/y5936e/ y5936e00.pdf.

Kelly, T., Yang, W., Chen, C. S., Reynolds, K. and He, J. 2008. Global burden of obesity in 2005 and projections to 2030. *International Journal of Obesity*, 32 (9): 1431–1437. Available at: http://dx.doi.org/10.1038/ijo.2008.102.

Kennedy, E. 1995. *The Healthy Eating Index*. US Department of Agriculture, Center for Nutrition Policy and Promotion, 34 pp. Available at: http://www.cnpp.usda.gov/ publications/hei/HEI89-90report.pdf.

Kesse, E. 2010. Typologies alimentaires pour mieux comprendre l'alimentation dans sa globalité. In P. Etiévant, F. Bellisle, J. Dallongeville *et al.*, eds. *Les comportements alimentaires. Quels en sont les determinants? Quelles actions, pour quels effets? Expertise scientifique collective*. Paris: INRA, pp. 29–38. Available at: http://www.inra.fr/content/download/25212/330233/version/1/file/ esco-inra-comportements-rapport-complet.pdf.

Key, T. J., Appleby, P. N. and Rosell, M. S. 2006. Health effects of vegetarian and vegan diets. *Proceedings of the Nutrition Society*, 65 (01): 35–41. Available at: http://dx.doi.org/10.1079/PNS2005481.

Kharas, H. 2010. The emerging middle class in developing countries. Working Paper no. 285. Paris: OECD Development Centre, 61 pp. Available at: http://www.oecd.org/dataoecd/12/52/44457738.pdf.

Kitinoja, L., Saran, S., Roy, S. K. and Kader, A. A. 2011. Postharvest technology for developing countries: challenges and opportunities in research, outreach and advocacy. *Journal of the Science of Food and Agriculture*, 91 (4): 597–603. Available at: http://dx.doi.org/10.1002/jsfa.4295.

Kling, M. and Hough, I. 2010. The American carbon foodprint: understanding your food's impact on climate change. Shelburne, VT: Brighter Planet, Inc., 17 pp. Available at: http://attachments.brighterplanet.com/press_items/local_copies/52/original/carbon_foodprint_wp.pdf?1271438613.

Kloppenburg, J., Jr, Hendrickson, J. and Stevenson, G. W. 1996. Coming in to the foodshed. *Agriculture and Human Values*, 13 (3): 33–42. Available at: http://dx.doi.org/10.1007/bf01538225.

Koellner, T. and Scholz, R. W. 2008. Assessment of land use impacts on the natural environment. Part 2: Generic characterization factors for local species diversity in central Europe. *International Journal of Life Cycle Assessment*, 13 (1): 32–48. Available at: http://dx.doi.org/10.1065/lca2006.12.292.2.

Koning, N. B. J., Van Ittersum, M. K., Becx, G. A. *et al.* 2008. Long-term global availability of food: continued abundance or new scarcity? *NJAS – Wageningen Journal of Life Sciences*, 55 (3): 229–292. Available at: http://dx.doi.org/10.1016/S1573-5214(08)80001-2.

Krugman, P. R. 1995. *Development, Geography, and Economic Theory*. Cambridge, MA: MIT Press, 117 pp.

Kruse, S. A., Flysjo, A., Kasperczyk, N. and Scholz, A. J. 2009. Socioeconomic indicators as a complement to life cycle assessment: an application to salmon production systems. *International Journal of Life Cycle Assessment*, 14 (1): 8–18. Available at: http://dx.doi.org/10.1007/s11367-008-0040-x.

Kumar, K. and Ayyappan, S. 1998. Current practises in integrated aquaculture, Working Paper, no. 5. The Integrated Aquaculture Research Planning Workshop, Purulia, India, March 1998. Institute of Aquaculture, University of Stirling, 25 pp. Available at: http://www.dfid.gov.uk/R4D/PDF/Outputs/NatResSys/R6759-Current-practices.pdf.

Lafay, L. C. and Volatier, J. L. C. 2009. Etude individuelle nationale des consommations alimentaires 2 (INCA 2), 2006–2007. Maisons-Alfort, France: AFSSA, 228 pp. Available at: http://www.afssa.fr/Documents/PASER-Ra-INCA2.pdf.

Lambert, C. 2009. *Les modalités de formation des prix alimentaires: du producteur au consommateur*. Paris: Conseil Économique, Social et Environnemental, 119 pp. Available at: http://www.conseil-economique-et-social.fr/rapport/rapsec/RS083920.pdf.

Lamine, C. 2008. Les intermittents du bio. Pour une sociologie pragmatique des choix alimentaires émergents. Paris: Quae-MSH (Natures sociales), 341 pp.

Lang, T. 2005. Food control or food democracy? Re-engaging nutrition with society and the environment. *Public Health Nutrition*, 8 (6a): 730–737. Available at: http://dx.doi.org/10.1079/PHN2005772.

Lang, T., Dibb, S. and Reddy, S. 2011. Looking backward, looking forward. Sustainability and UK food policy 2000–2011. London: Sustainable Development Commission, 60 pp. Available at: http://www.sd-commission.org.uk/data/files/publications/FoodPolicy10_Report_final_w.pdf.

Le Nechet, R., Michaud, M., Legrain, M. *et al.* 2006. *2020: que mangerons-nous ? Enjeux pour les productions agricoles normandes.* Normandy, France: CRA Normandie, 8 pp. Available at: http://www.normandie.chambagri.fr/2020/documents/C00.pdf.

Leclerc, A., Chastang, J. F., Menvielle, G. and Luce, D. 2006. Socioeconomic inequalities in premature mortality in France: have they widened in recent decades? *Social Science & Medicine*, 62 (8): 2035–2045. Available at: http://dx.doi.org/10.1016/j.socscimed.2005.08.031.

Ledikwe, J. H., Blanck, H. M., Khan, L. K. *et al.* 2006. Low-energy-density diets are associated with high diet quality in adults in the United States. *Journal of the American Dietetic Association*, 106 (8): 1172–1180. Available at: http://dx.doi.org/10.1016/j.socscimed.2005.08.031.

Lee-Smith, D. 2010. Cities feeding people: an update on urban agriculture in equatorial Africa. *Environment and Urbanization*, 22 (2): 483–499. Available at: http://eau.sagepub.com/content/22/2/483.abstract.

Lemay, J.-F., Favreau, L. and Maldidier, C. 2010. *Commerce équitable: les défis de la solidarité dans les échanges internationaux.* Quebec, Canada: Presses de l'Université du Québec (*Collection Initiatives*), 170 pp.

Leonard, W. R., Snodgrass, J. J. and Robertson, M. L. 2010. Evolutionary perspectives on fat ingestion and metabolism in humans. *Fat Detection: Taste, Texture, and Post Ingestive Effects. Frontiers in Neuroscience.* Boca Raton, FL: CRC Press. Available at: http://www.ncbi.nlm.nih.gov/books/NBK53561/.

Levett, R. 1998. Footprinting: a great step forward, but tread carefully. A response to Mathis Wackernagel. *Local Environment: The International Journal of Justice and Sustainability*, 3 (1): 67–74. Available at: http://www.informaworld.com/10.1080/13549839808725545.

Liang, L. *et al.* 1993. China's post-harvest grain losses and the means of their reduction and elimination. Jingji dili (Econ. Geogr.), 1: 92–96.

Lloyd-Williams, F., O'Flaherty, M., Mwatsama, M. *et al.* 2007. Estimating the cardiovascular mortality burden attributable to the European Common Agricultural Policy on dietary saturated fats. *Bulletin of the World Health Organization*, 86 (7): 535–542. Available at: http://www.who.int/entity/bulletin/volumes/86/7/08-053728.pdf.

Lobstein, T. 2002. Food policies: a threat to health? *Proceedings of the Nutrition Society*, 61 (04): 579–585. Available at: http://dx.doi.org/10.1079/PNS2002192.

Lock, K., Smith, R. D., Dangour, A. D. *et al.* 2010. Health, agricultural, and economic effects of adoption of healthy diet recommendations. *Lancet*, 376 (9753): 1699–1709. Available at: http://dx.doi.org/10.1016/S0140-6736(10)61352-9.

London Development Agency 2006. *Healthy and Sustainable Food for London*. London: Mayor of London, 139 pp. Available at: http://vancouver.ca/commsvcs/socialplanning/initiatives/foodpolicy/tools/pdf/London_Food_Policy.PDF.

Lorne, D. and Bonnet, J. F. 2009. Eau et biocarburants à l'horizon 2030. Impacts sur l'eau du développement des biocarburants en France à l'horizon 2030. *Les cahiers du CLIP*, 19: 1–98. Available at: http://www.iddri.org/Publications/Les-cahiers-du-CLIP/clip_19.pdf.

Lowe, P., Phillipson, J. and Lee, R. P. 2008. Socio-technical innovation for sustainable food chains: roles for social science. *Trends in Food Science & Technology*, 19 (5): 226–233. Available at: http://dx.doi.org/10.1016/j.tifs.2007.11.005.

Lundqvist, J., de Fraiture, C. and Molden, D. 2008. Saving water: from field to fork – curbing losses and wastage in the food chain. Stockholm: The Stockholm International Water Institute (SIWI), 36 pp. Available at: http://dx.doi.org/10.1016/j.tifs.2007.11.005.

Lutz, W., Sanderson, W. and Scherbov, S. 2001. The end of world population growth. *Nature*, 412 (6846): 543–545. Available at: http://dx.doi.org/10.1038/35087589.

Lutz, W., Sanderson, W. and Scherbov, S. 2008. The coming acceleration of global population ageing. *Nature*, 451 (7179): 716–719. Available at: http://dx.doi.org/10.1038/nature06516.

Mackenbach, J. P., Stirbu, I., Roskam, A.-J. R. *et al.* 2008. Socioeconomic inequalities in health in 22 European countries. *New England Journal of Medicine*, 358 (23): 2468–2481. Available at: http://www.nejm.org/doi/full/10.1056/NEJMsa0707519.

Madden, J. P., Goodman, S. J. and Guthrie, H. A. 1976. Validity of the 24-hr. recall. Analysis of data obtained from elderly subjects. *Journal of the American Dietetic Association*, 68 (2): 143–147.

Maertens, M. and Swinnen, J. F. M. 2009. Trade, standards, and poverty: evidence from Senegal. *World Development*, 37 (1): 161–178. Available at: http://dx.doi.org/10.1016/j.worlddev.2008.04.006.

Maillot, M., Darmon, N., Darmon, M., Lafay, L. and Drewnowskis, A. 2007. Nutrient-dense food groups have high energy costs: an econometric approach to nutrient profiling. *Journal of Nutrition*, 137 (7): 1815–1820. Available at: http://jn.nutrition.org/content/137/7/1815.full.pdf+html.

Maillot, M., Vieux, F., Amiot, M. J. and Darmon, N. 2010. Individual diet modeling translates nutrient recommendations into realistic and individual-specific food choices. *American Journal of Clinical Nutrition*, 91 (2): 421–430. Available at: http://www.ajcn.org/content/91/2/421.full.pdf+html.

Maillot, M., Vieux, F., Ferguson, E. F. *et al.* 2009. To meet nutrient recommendations, most French adults need to expand their habitual food repertoire. *Journal of Nutrition*, 139 (9): 1721–1727. Available at: http://www.ajcn.org/content/91/2/421.full.pdf+html.

Mainguy, P. 1989. *La qualité dans l'agro-alimentaire*. Paris: Ministère de l'agriculture et le secrétariat d'état à la consommation, 58 pp.

Maire, B. and Mejean, C. 2008. Que savons-nous de l'alimentation des migrants ? *Lettre scientifique de l'IFN*, 129: 1–10. Available at: http://www.ifn.asso.fr/publications/lettres/pdf_let-scien-n-129.pdf.

Malassis, L. 1996. Les trois ages de l'alimentaire. *Agroalimentaria*, 96 (2): 3–5. Available at: http://www.saber.ula.ve/bitstream/123456789/17732/1/articulo2_1.pdf.

Manalili, N. M., Dorado, M. A. and Van Otterdijk, R. 2011. *Appropriate Food Packaging Solutions for Developing Countries*. Rome: FAO, 28 pp. Available at: http://www.messe-duesseldorf.de/save-food/doc/Appropriate_Packaging_Solutions.pdf.

Mann, N. 2000. Dietary lean red meat and human evolution. *European Journal of Nutrition*, 39 (2): 71–79. Available at: http://dx.doi.org/10.1007/s003940050005.

Mannie, M. D. 2010. Autoimmunity and asthma: the dirt on the hygiene hypothesis. *Self/Nonself*, 1 (2): 123–128. Available at: http://www.landesbioscience.com/journals/selfnonself/article/11550/.

Marlow, H. J., Hayes, W. K., Soret, S. *et al.* 2009. Diet and the environment: does what you eat matter? *American Journal of Clinical Nutrition*, 89 (5): 1699S–1703S. Available at: http://dx.doi.org/10.3945/ajcn.2009.26736Z.

Marshall, D. W. and Anderson, A. S. 2002. Proper meals in transition: young married couples on the nature of eating together. *Appetite*, 39 (3): 193–206. Available at: http://dx.doi.org/10.1006/appe.2002.0507.

Martin, W. and Anderson, K. 2011. Export restrictions and price insulation during commodity price booms. Policy Research Working Paper, WPS no. 5645. Washington DC: The World Bank, 21 pp. Available at: http://www-wds.worldbank.org/external/default/WDSContentServer/IW3P/IB/2011/05/02/000158349_20110502082850/Rendered/PDF/WPS5645.pdf.

Martinez, S., Hand, M., Da Pra, M. *et al.* 2010. *Local Food Systems: Concepts, Impacts, and Issues*. Washington DC: USDA–ERS, 87 pp. Available at: http://www.ers.usda.gov/Publications/ERR97/ERR97.pdf.

Masset, E. 2011. A review of hunger indices and methods to monitor country commitment to fighting hunger. *Food Policy*, 36 (Supplement 1): S102–S108. Available at: http://dx.doi.org/10.1016/j.foodpol.2010.11.007.

Masullo, A. and Régnier, F. 2009. Obésité, goûts et consommation. Intégration des normes d'alimentation et appartenance sociale. *Revue Française de Sociologie*, 50(4): 747–773. Available at: http://www.cairn.info/revue-francaise-de-sociologie-2009-4-p-747.htm.

Mazoyer, M. and Roudart, L. 2002. *Histoire des agricultures du monde: du néolithique à la crise contemporaine*. Paris: Éditions du Seuil, 705 pp.

Mazzocchi, M., Brasili, C. and Sandri, E., 2008. Trends in dietary patterns and compliance with World Health Organization recommendations: a cross-country analysis. *Public Health Nutrition*, 11 (05): 535–540. Available at: http://dx.doi.org/10.1017/S1368980007000900.

Mazzocchi, M., Traill, B. and Shogren, J. F. 2009. *Fat Economics: Nutrition, Health, and Economic Policy*. Oxford: Oxford University Press, 208 pp.

McIntyre, B. D., Herren, H. R., Wakhungu, J. and Watson, R. T. 2009. Agriculture at a crossroads: global report. Washington DC: International Assessment of Agricultural Knowledge, Science, and Technology for Development (IAASTD), 590 pp. Available at: http://www.agassessment.org/reports/IAASTD/EN/Agriculture at a Crossroads_Global Report (English).pdf.

McMichael, A. J., Powles, J. W., Butler, C. D. and Uauy, R. 2007. Food, livestock production, energy, climate change, and health. *The Lancet*, 370 (9594): 1253–1263. Available at: http://dx.doi.org/10.1016/S0140-6736(07)61256-2.

Melo, M. T., Nickel, S. and Saldanha-da-Gama, F. 2009. Facility location and supply chain management: a review. *European Journal of Operational Research*, 196 (2): 401–412. Available at: http://dx.doi.org/10.1016/j.ejor.2008.05.007.

Meng, X. and Qian, N. 2009. The long term consequences of famine on survivors: evidence from a unique natural experiment using China's great famine. NBER Working Paper, 49 pp. Available at: http://www.nber.org/papers/w14917.

Meuwissen, M. P. M., Velthuis, A. G. J., Henk, H. and Huirne, R. B. M. 2003. Traceability and certification in meat supply chains. *Journal of Agribusiness*, 21 (2). Available at: http://econpapers.repec.org/RePEc:ags:jloagb:14666.

Micha, R., Wallace, S. K. and Mozaffarian, D. 2010. Red and processed meat consumption and risk of incident coronary heart disease, stroke, and diabetes mellitus: a systematic review and meta-analysis. *Circulation*, 121 (21): 2271–2283. Available at: http://circ.ahajournals.org/cgi/content/abstract/121/21/2271.

Mila i Canals, L., Cowell, S. J., Sim, S. and Basson, L. 2007. Comparing domestic versus imported apples: a focus on energy use. *Environmental Science and Pollution Research*, 14 (5): 338–344. Available at: http://dx.doi.org/10.1065/espr2007.04.412.

Millet, M. and Thin, D. 2005. Le temps des familles populaires à l'épreuve de la précarité. *Lien social et Politiques*, 54: 153–162. Available at: http://www.erudit.org/revue/LSP/2005/v/n54/012867ar.pdf.

Millward, D. J. and Garnett, T. 2010. Food and the planet: nutritional dilemmas of greenhouse gas emission reductions through reduced intakes of meat and dairy foods. *Proceedings of the Nutrition Society*, 69 (01): 103–118. Available at: http://dx.doi.org/10.1017/S0029665109991868.

Minegishi, S. and Thiel, D. 2000. System dynamics modeling and simulation of a particular food supply chain. *Simulation Practice and Theory*, 8 (5): 321–339. Available at: http://dx.doi.org/10.1016/s0928-4869(00)00026-4.

Minten, B., Randrianarison, L. and Swinnen, J. F. M. 2009. Global retail chains and poor farmers: evidence from Madagascar. *World Development*, 37 (11): 1728–1741. Available at: http://dx.doi.org/10.1016/j.worlddev.2008.08.024.

Moneret-Vautrin, D. A. 2008. Épidémiologie de l'allergie alimentaire. *Revue Française d'Allergologie et d'Immunologie Clinique*, 48 (3): 171–178. Available at: http://dx.doi.org/10.1016/j.allerg.2008.01.018.

Morgan, K. 2010. Local and green, global and fair: the ethical foodscape and the politics of care. *Environment and Planning A*, 42 (8): 1852–1867. Available at: http://www.envplan.com/abstract.cgi?id=a42364.

Morgan, K. (ed.) 2009. Feeding the city: the challenge of urban food planning. *International Planning Studies*, 14 (4): 341–348 [Special issue].

Morgan, K. and Sonnino, R. 2010. The urban foodscape: world cities and the new food equation. *Cambridge Journal of Regions, Economy and Society*, 3 (2): 209–224. Available at: http://cjres.oxfordjournals.org/cgi/content/abstract/3/2/209.

Moustier, P., Chaleard, J. L. and Leplaideur, A. 2002. L'approvisionnement vivrier des villes en Guinée: entre fragilité et dynamisme. *Autrepart*, 23: 5–23.

Muchnik, J., Sanz Cañada, J. and Torres Salcido, G. 2008. Systèmes agroalimentaires localisés: état des recherches et perspectives. *Cahiers Agriculture*, 17 (6): 513–519. Available at: http://www.john-libbey-eurotext.fr/fr/revues/agro_biotech/agr/e-docs/00/04/44/20/article.phtml.

Müller-Lindenlauf, M., Deittert, C. and Köpke, U. 2010. Assessment of environmental effects, animal welfare and milk quality among organic dairy farms. *Livestock Science*, 128 (1–3): 140–148. Available at: http://dx.doi.org/10.1016/j.livsci.2009.11.013.

Muñoz, I., Milà i Canals, L. and Clift, R. 2008. Consider a spherical man. *Journal of Industrial Ecology*, 12 (4): 521–538. Available at: http://dx.doi.org/10.1111/j.1530-9290.2008.00060.x.

Murthy, D. S., Gajanana, T. M., Sudha, M. and Dakshinamoorthy, V. 2009. Marketing and post-harvest losses in fruits: its implications on availability and economy. *Indian Journal of Agricultural Economics*, 64 (2): 259–275. Available at: http://econpapers.repec.org/RePEc:ces:ceswps:_2994.

Nadvi, K. and Waltring, F. 2003. Making sense of global standards. In H. Schmitz, ed. *Local Enterprises in the Global Economy: Issues of Governance and Upgrading*. Cheltenham, UK: Edward Elgar, pp. 53–94.

NASA-Goddard Institute for Space Studies 2012. GISS surface temperature analysis. GISS, New York. Available at http://data.giss.nasa.gov/gistemp/graphs_v3/, accessed 11 July 2011.

Neelsen, S. and Stratmann, T. 2010. Effects of prenatal and early life malnutrition: evidence from the Greek famine. Working Paper Series, CESifo, 35 p. Available at: http://econpapers.repec.org/RePEc:ces:ceswps:_2994.

Neilson, J. 2008. Global private regulation and value-chain restructuring in Indonesian smallholder coffee systems. *World Development (Oxford)*, 36 (9): 1607–1622. Available at: http://dx.doi.org/10.1016/j.worlddev.2007.09.005.

Neilson, J. and Pritchard, B. 2009. Introduction. In J. Neilson and B. Pritchard, eds, *Value Chain Struggles: Institutions and Governance in the Plantation Districts of South India*. Oxford: Wiley-Blackwell. Available at: http://dx.doi.org/10.1002/9781444308723.ch1.

Nellemann, C., MacDevette, M., Manders, T. *et al.* 2009. *The Environmental Food Crisis: The Environment's Role in Averting Future Food Crisis*. Nairobi: United Nations Environment Programme (UNEP), 104 pp. Available at: http://www.grida.no/files/publications/FoodCrisis_lores.pdf.

Newbery, D. M. 1989. The theory of food price stabilisation. *The Economic Journal*, 99 (398): 1065–1082. Available at: http://www.jstor.org/stable/2234088.

Newsom, G. 2009. Executive Directive 09–03: Healthy and Sustainable Food for San Francisco. San Francisco: Office of the Mayor City & County of San Francisco, 5 pp. Available at: http://www.sfgov3.org/ftp/uploadedfiles/sffood/policy_reports/MayorNewsomExecutiveDirectiveonHealthySustainableFood.pdf.

Nichele, V. 2003. Health information and food demand in France. In W. S. Chern and K. Rickertson, eds, *Health, Nutrition and Food Demand*. Wallingford, UK: CABI Publishing, pp. 131–151.

ObEpi 2009. *Enquête épidémiologique nationale sur le surpoids et l'obésité. ObÉpi* 2009 Neuilly-sur-Seine, France: Inserm/TNS Healthcare/Laboratoire Roche, 56 pp. Available at: http://www.roche.fr/gear/newcontents/servlet/staticfilesServlet?type=data&communityId=re719001&id=static/attachedfile/re7300002/re72700003/AttachedFile_10160.pdf.

Odum, H. T. 1996. *Environmental Accounting: Emergy and Environmental Decision Making.* New York: Wiley, 370 pp.

OECD 1997. *OECD Environmental Performance Reviews: A Practical Introduction.* Paris: OECD, 60 pp. Available at: http://www.fao.org/ag/againfo/programmes/fr/lead/toolbox/Refer/gd9735.pdf.

OECD 1999. *An Assessment of the Costs for International Trade in Meeting Regulatory Requirements.* Paris: OECD, 118 pp. Available at: http://www.oecd.org/dataoecd/33/14/1955269.pdf.

OECD 2008a. *Highlights of the International Transport Forum 2008: Transport and Energy The Challenge of Climate Change.* Paris: OECD, 72 pp.

OECD 2008b. *Croissance et inégalités.* Paris: OECD, 10 pp. Available at: http://www.oecd.org/dataoecd/48/9/41530189.pdf.

Okada, H., Kuhn, C., Feillet, H. and Bach, J. F. 2010. The 'hygiene hypothesis' for autoimmune and allergic diseases: an update. *Clinical & Experimental Immunology,* 160 (1): 1–9.

ONPES 2010. Bilan de 10 ans d'observation de la pauvreté et de l'exclusion sociale à l'heure de la crise. Rapport 2009–2010. Paris: ONPES, 164 pp. Available at: http://www.fao.org/docrep/005/ac911e/ac911e00.htm.

Osunde, Z. D. 2008. Minimizing postharvest losses in yam (*Dioscorea* spp.): treatments and techniques. In G. L. Robertson and J. R. Lupien, eds, *Using Food Science and Technology to Improve Nutrition and Promote National Development.* Toronto, Canada: International Union of Food Science & Technology, Chapter 12. Available at: http://www.iufost.org/publications/books/documents/Osundechapter12.pdf.

Pachauri, R. K. and Reisinger, A. 2007. Bilan 2007 des changements climatiques. Rapport de synthèse. Geneva: GIEC, 103 pp. Available at: http://www.ipcc.ch/pdf/assessment-report/ar4/syr/ar4_syr_fr.pdf.

Padilla, M. and Abis, S. 2007. La grande distribution au Maghreb. *Afkar/idées,* 13 (Printemps 2007): 68–71. Available at: http://www.afkar-ideas.com/wp-content/uploads/files/3-13-20-fr.pdf.

Paillard, S. C., Treyer, S. C. and Dorin, B. C. 2010. *Agrimonde: Scenarios and Challenges for Feeding the World in 2050.* Paris: Quae (Matière à débattre et décider), 295 pp.

Parfitt, J., Barthel, M. and Macnaughton, S. 2010. Food waste within food supply chains: quantification and potential for change to 2050. *Philosophical Transactions of the Royal Society B: Biological Sciences,* 365 (1554): 3065–3081. Available at: http://rstb.royalsocietypublishing.org/content/365/1554/3065.full.pdf.

Pattberg, P. 2005. The institutionalization of private governance: how business and nonprofit organizations agree on transnational rules. *Governance,* 18 (4): 589–610. Available at: http://www.glogov.org/images/doc/Pattberg 2005.pdf.

Patterson, R. E., Haines, P. S. and Popkin, B. M. 1994. Diet quality index: capturing a multidimensional behavior. *Journal of the American Dietetic Association*, 94 (1): 57–64. Available at: http://dx.doi.org/10.1016/0002-8223(94)92042-7.

Périssé, J., Sizaret, F. and François, P. 1969. Effet du revenu sur la structure de la ration alimentaire. *Bulletin de Nutrition FAO*, 7 (3): 1–10.

Perrier-Cornet, P. 2009. Les systèmes agroalimentaires localisés sont-ils ancrés localement ? Un bilan de la littérature contemporaine sur les Syal. In F. Aubert, V. Piveteau and B. Schmitt, eds, *Politiques agricoles et territoires*. Versailles, France: Editions Quae, pp. 49–68.

Pesqueux, Y. 2006. Pour une évaluation critique de la théorie des parties prenantes. In M. Bonnafous-Boucher and Y. Pesqueux, eds, *Décider avec les parties prenantes. Approches d'une nouvelle théorie de la société civile*. Paris: La Découverte, pp. 19–40.

Phillipson, J. and Lowe, P. 2008. Towards sustainable food chains: harnessing the social and natural sciences. *Trends in Food Science & Technology*, 19 (5): 224–225. Available at: http://dx.doi.org/10.1016/j.tifs.2008.01.001.

Pimentel, D., Hurd, L. E., Bellotti, A. C. *et al.* 1973. Food production and the energy crisis. *Science*, 182 (4111): 443–449. Available at: http://dx.doi.org/10.1126/science.182.4111.443.

Pingali, P. L., Bigot, Y. and Binswanger, H. P. 1987. *Agricultural Mechanization and the Evolution of Farming Systems in Sub-Saharan Africa*. Baltimore, MD: Johns Hopkins University Press, 216 pp. Available at: http://dx.doi.org/10.1016/0304-3878(89)90039-4.

PIPAME 2009. *L'impact des technologies de l'information sur la logistique*. Paris: Ministère de l'Economie, des finances et de l'industrie. 56 pp. Available at: http://www.industrie.gouv.fr/p3e/etudes/logistique_tic/etudes8.php.

PNUE 2007. GEO-4 Environment for Development. Nairobi: PNUE, 540 pp. Available at: http://www.unep.org/geo/GEO4/report/GEO-4_Report_Full_FR.pdf.

Popkin, B. M. 1999. Urbanization, lifestyle changes and the nutrition transition. *World Development*, 27 (11): 1905–1916. Available at: http://dx.doi.org/10.1016/S0305-750X(99)00094-7.

Popkin, B. M. 2006. Global nutrition dynamics: the world is shifting rapidly toward a diet linked with noncommunicable diseases. *American Journal of Clinical Nutrition*, 84 (2): 289–298. Available at: http://www.ajcn.org/content/84/2/289.full.

Popkin, B. M. 2011. Contemporary nutritional transition: determinants of diet and its impact on body composition. *Proceedings of the Nutrition Society*, 70 (01): 82–91. Available at: http://dx.doi.org/10.1017/S0029665110003903.

Porter, M. E. 1996. What is strategy? *Harvard Business Review*, (November–December): 61–78. Available at: http://www.ipocongress.ru/download/guide/article/what_is_strategy.pdf.

Poulain, J.-P. 2009. *Sociologie de l'obésité*. Paris: PUF (Sciences sociales et sociétés), 386 pp.

Prättälä, R. 2003. Dietary changes in Finland: success stories and future challenges. *Appetite*, 41 (3): 245–249. Available at: http://www.ipocongress.ru/download/guide/article/what_is_strategy.pdf.

Premier Ministre and Centre d'analyse stratégique 2009. France-2025 C – Groupe 3: Technologies et vie quotidienne. In: d'Aubert, F., ed. *France 2025. Diagnostic stratégique*. Paris: Premier Ministre, 57 pp. Available at: http://temis.documentation. equipement.gouv.fr/documents/Temis/0063/Temis-0063309/17680_3.pdf.

Prentice, C. W. C., Harrison, S. P., Leemans, R., Monserud, R. A. and Solomon, A. M. 1992. A global biome model based on plant physiology and dominance, soil properties and climate. *Journal of Biogeography*, 19: 117–134. Available at: http:// www.jstor.org/stable/2845499.

Pretty, J., Sutherland, W. J., Ashby, J. *et al.* 2010. The top 100 questions of importance to the future of global agriculture. *International Journal of Agricultural Sustainability*, 8 (4): 219–236. Available at: http://dx.doi.org/10.3763/ijas.2010.0534.

Pretty, J. N. 2008. *Sustainable Agriculture and Food*. London: Earthscan, 600 pp.

Pretty, J. N., Ball, A. S., Lang, T. and Morison, J. I. L. 2005. Farm costs and food miles: an assessment of the full cost of the UK weekly food basket. *Food Policy*, 30 (1): 1–19. Available at: http://dx.doi.org/10.1016/j.foodpol.2005.02.001.

Princen, T. 1997. The shading and distancing of commerce: when internalization is not enough. *Ecological Economics*, 20 (3): 235–253. Available at: http://dx.doi.rg/ 10.1016/S0921-8009(96)00085-7.

Rajagopal, D. and Zilberman, D. 2007. Review of environmental, economic and policy aspects of biofuels. Policy Research Working Paper Series, WPS no. 4341. Washington DC: The World Bank, 107 pp. Available at: http://ideas.repec.org/p/ wbk/wbrwps/4341.html.

Rastoin, J. L., Ghersi, G. and De Schutter, O. 2010. *Le système alimentaire mondial: concepts et méthodes, analyses et dynamiques*. Versailles, France: Editions Quae (*Synthèses*), 565 pp.

Ravallion, M. 2010. The developing world's bulging (but vulnerable) middle class. *World Development*, 38 (4): 445–454. Available at: http://dx.doi.org/10.1016/j. worlddev.2009.11.007.

Ravallion, M., Chen, S. and Sangraula, P. 2007. New evidence on the urbanization of global poverty. *Population and Development Review*, 33 (4): 667–701. Available at: http://dx.doi.org/10.1111/j.1728-4457.2007.00193.x.

Raynolds, L. T. 2000. Re-embedding global agriculture: the international organic and fair trade movements. *Agriculture and Human Values*, 17 (3): 297–309. Available at: http://dx.doi.org/10.1023/a:1007608805843.

Raynolds, L. T., Murray, D. and Heller, A. 2007. Regulating sustainability in the coffee sector: a comparative analysis of third-party environmental and social certification initiatives. *Agriculture and Human Values*, 24 (2): 147–163. Available at: http://springerlink.metapress.com/link.asp?id=402841.

REAP 2010. Services on Monitoring Retailers' REAP commitments (Retailers' Environmental Action Programme). Consortium ESWI, Expert Team to Support Waste Implementation, 158 pp. Available at: http://ec.europa.eu/environment/ industry/retail/pdf/report_monitoring2009.pdf.

Reardon, T. and Gulati, A. 2008. The rise of supermarkets and their development implications: international experience relevant for India. Washington DC: IFPRI,

Michigan State University, 60 pp. Available at: http://www.ifpri.org/sites/default/files/publications/ifpridp00752.pdf.

Reardon, T., Henson, S. and Berdegue, J. 2007. 'Proactive fast-tracking' diffusion of supermarkets in developing countries: implications for market institutions and trade. *Journal of Economic Geography*, 7 (4): 399–431. Available at: http://users.unimi.it/olper/pdf/materiale-integrativo/Reardon-Henson-Berdeue 2007JEG.pdf.

Recours, F. and Hébel, P. 2006. Les populations modestes ont-elles une alimentation déséquilibrée? Paris: CREDOC (Cahiers de Recherche), 113 pp. Available at: http://www.credoc.fr/pdf/Rech/C232.pdf.

Regmi, A. and Unnevehr, L. 2006. Are diets converging globally? A comparison of trends across selected countries. *Journal of Food Distribution Research*, 37 (1): 14–21.

Regmi, A., Takeshima, H. and Unnevehr, L. J. 2008. *Convergence in Global Food Demand and Delivery*. Economic Research Report, no. 56, Washington DC: USDA, 33 pp. Available at: http://www.ers.usda.gov/publications/err56/err56.pdf.

Régnier, F. 2006. Manger hors norme, respecter les normes: le plaisir de l'exotisme culinaire. *Journal des Anthropologues*, 106–107 (169–187). Available at: http://jda.revues.org/1315.

Reijnders, L. and Soret, S. 2003. Quantification of the environmental impact of different dietary protein choices. *American Journal of Clinical Nutrition*, 78 (3): 664S–668S. Available at: http://www.ajcn.org/content/78/3/664S.abstract.

RESEDA 2005. *Gisements des coproduits, sous-produits et déchets des industries alimentaires*. RÉseau des organisations professionnelles et interprofessionnelles pour la SÉcurité et la qualité des Denrées Animales, 119 pp.

Rey-Valette, H., Clément, O., Aubin, J. *et al.* 2008. *Guide de co-construction d'indicateurs de développement durable en aquaculture* Montpellier, France: Cirad/Ifremer/INRA/IRD/Université Montpellier 1, projet « EVAD » (Evaluation de la durabilité des systèmes aquacoles), 144 pp.

Risku-Norja, H., Hietala, R., Virtanen, H., Ketomäki, H. and Helenius, J. 2008. Localisation of primary food production in Finland: production potential and environmental impacts of food consumption patterns. *Agricultural and Food Science*, 17 (2): 127–145. Available at: http://dx.doi.org/10.2137/145960608785328233.

Rizet, C. and Keita, B. 2005. Chaînes logistiques et consommation d'énergie. Cas du yaourt et du jean. Paris: INRETS-ADEME, 81 pp. Available at: http://www.inrets.fr/fileadmin/ur/dest/PDF/Journee_Supply-Chain/Rapport-INRETS-Supply_chain_Energie-juin-05.pdf.

Rosa, H. 2010. *Accélération: Une critique sociale du temps*. Paris: La Découverte, 480 pp.

Rossier, D. and Gaillard, G. 2004. Ökobilanzierung des Landwirtschaftsbetriebs: Methode und Anwendung in 50 Landwirtschaftsbetrieben. *FALSchriftenreihe no.53. Forschungsanstalt für Agrarökologie und Landbau (FAL), Zürich, Switzerland*.

Rovillé-Sausse, F. 1999. Evolution en 20 ans de la corpulence des enfants de 0 à 4 ans issus de l'immigration maghrébine. *Revue Epidémiologique de Santé Publique*, 47 (1): 37–44.

Rowe, G. and Wright, G. 1999. The Delphi technique as a forecasting tool: issues and analysis. *International Journal of Forecasting*, 15 (4): 353–375. Available at: http://dx.doi.org/10.1016/s0169-2070(99)00018-7.

Roy, B. 1968. Classement et choix en présence de points de vue multiples (la méthode ELECTRE). *Revue d'Informatique et de Recherche Opérationelle (RIRO)*, 8: 57–75.

Ruttan, V. W. 2002. Productivity growth in world agriculture: sources and constraints. *The Journal of Economic Perspectives*, 16 (4): 161–184. Available at: http://www.jstor.org/stable/3216919.

Satterthwaite, D., McGranahan, G. and Tacoli, C. 2010. Urbanization and its implications for food and farming. *Philosophical Transactions of the Royal Society B: Biological Sciences*, 365 (1554): 2809–2820. Available at: http://pubmedcentralcanada.ca/picrender.cgi?accid=PMC2935117&blobtype=pdf.

Sauvant, A. 2002. Volume et partage modal du transport de marchandises en France de 1845 à nos jours. *Notes de synthèse du Service économique et statistique*, 140: 18–26. Available at: http://temis.documentation.equipement.gouv.fr/documents/Temis/0041/Temis-0041905/NS_140_4.pdf.

Savin, J.-M. 2000. L'évolution des distances moyennes de transports des marchandises. *Notes de synthèse du Service économique et statistique*, 129: 18–26. Available at: http://temis.documentation.equipement.gouv.fr/documents/Temis/0034/Temis-0034954/NS_129_4.pdf.

Schaffnit-Chatterjee, C. 2009. *The Global Food Equation: Food Security in an Environment of Increasing Scarcity*. Frankfurt, Germany: Deutsche Bank Research, 38 pp. Available at: http://www.dbresearch.com/PROD/DBR_INTERNET_EN-PROD/PROD0000000000247631.pdf.

Schlich, E. H. and Fleissner, U. 2005. The ecology of scale: assessment of regional energy turnover and comparison with global food. *International Journal of Life Cycle Assessment*, 10 (3): 219–223. Available at: http://dx.doi.org/10.1065/lca2004.09.180.9.

Schmidhuber, J. 2004. The growing global obesity problem: some policy options to address it. *FAO food and nutrition paper* (Globalization of food systems in developing countries: impact on food security and nutrition) 33: 81–97. Available at: ftp://ftp.fao.org/docrep/fao/007/y5736e/y5736e00.pdf.

Schmidhuber, J. and Shetty, P. 2005. The nutrition transition to 2030. Why developing countries are likely to bear the major burden. *Food Economics – Acta Agriculturae Scandinavica, Section C*, 2 (3): 150–166. Available at: http://www.informaworld.com/10.1080/16507540500534812.

Schmidhuber, J. and Traill, W. B. 2006. The changing structure of diets in the European Union in relation to healthy eating guidelines. *Public Health Nutrition*, 9 (05): 584–595. Available at: http://dx.doi.org/10.1079/PHN2005844.

Schulten, G. G. M. 1982. Post-harvest losses in tropical Africa and their prevention. *Food and Nutrition Bulletin*, 4 (2): 2–9. Available at: http://www.greenstone.org/greenstone3/nzdl?a=d&d=HASH5d7120ffe30226d142f3f7.2.np&c=hdl&sib=1&dt=&ec=&et=&p.a=b&p.s=ClassifierBrowse&p.sa=.

Schultz, I. and Stieß, I. 2008. Linking sustainable consumption to everyday life. A social-ecological approach to consumption research. In A. Tukker, M. Charter and C. Vezzoli, eds, *Perspectives on Radical Changes to Sustainable Consumption and Production. System Innovation for Sustainability*. Sheffield, UK: Greenleaf Publishing Ltd., pp. 288–300.

Scoones, I. 2009. Livelihoods perspectives and rural development. *Journal of Peasant Studies*, 36 (1): 171–196. Available at: http://www.tandf.co.uk/journals/pdf/papers/FJPS_36_1_2009.pdf.

Segerlund, L. 2010. *Making Corporate Social Responsibility a Global Concern: Norm Construction in a Globalizing World*. Non-state Actors in International Law, Politics and Governance Series. Farnham, UK: Ashgate, 208 pp. Available at: http://www.jstor.org/pss/1833080.

Seifu, S., Asrat, A., Argaw, A. *et al.* 2007. *Cooperative Unions and 'Sustainable' Coffee Initiatives in Ethiopia: Opportunities and Challenges*. Association Scientifique Internationale du Café (ASIC), pp. 473–486.

Sermet, C. 2006. Evaluation du coût associé à l'obésité en France. *La Presse Médicale* 36 (6): 832–840. Available at: http://www.irdes.fr/EspacePresse/QuestionsA/QuestionsAJuil07.html.

Short, F. 2006. *Kitchen Secrets, The Meaning of Cooking in Everyday Life*. London: Berg Publishers, 224 pp.

Sieber, J. 2006. WEAP Water Evaluation and Planning System *iEMSs 2006: 3rd Biennial meeting of the International Environmental Modelling and Software Society*. Burlington, VT, 9–13 July 2006, 6 pp. Available at: http://www.iemss.org/iemss2006/papers/w5/WEAP Desc.pdf.

Singano, C. D., Nkhata, B. T. and Mhango, V. 2008a. National annual report on larger grain borer monitoring and *Teretrius nigrescens* rearing and releases in Malawi. Plant Protection Progress Report for the 2007/2008 season, presented at the Department of Agricultural Research Services Planning and Review Meeting, Andrews Hotel, Mangochi, 14–20 September 2008, pp. 1–8.

Singano, C. D., Phiri, T., Nkhata, B. T. and Mhango, V. 2008b. National agricultural produce inspection services annual technical report for the period July 2007–June 2008. Plant Protection Progress Report for the 2007/2008 season, presented at the Department of Agricultural Research Services Planning and Review Meeting, Andrews Hotel, Mangochi, 14–20 September 2008, pp. 9–22.

Sirieix, L., Alessandrin, A. and Persillet, V. 2006. Motivations and values: a means-end chain study of French consumers. In G. C. Holt and M. Reed, eds, *Sociological Perspectives of Organic Agriculture: From Pioneer to Policy*. Wallingford, UK: CABI, pp. 70–87. Available at: http://books.google.fr/books?id=8kV06YS9qMgC&printsec=frontcover&dq=Sociological+perspectives+of+organic+agriculture:+from+pioneer+to+policy&hl=fr&ei=VOORTcnzDYzBswbq6ojQBg&sa=X&oi=book_result&ct=result&resnum=1&ved=0CC4Q6AEwAA#v=onepage&q&f=false.

SITA France and Banque alimentaire du Bas-Rhin 2011. Evaluation de l'impact sur l'effet de serre de l'action de lutte contre le gaspillage de la Banque Alimentaire du Bas-Rhin. Final Report, April 2011. Paris: SITA France, 19 pp. Available at: http://www.banquealimentaire.org/sites/default/files/etude_ffba_impact_ges_du_gaspillage_alimentaire_vf_avril_2011.pdf.

Sjauw-Koen-Fa, A. 2009. Sustainability and security of the global food supply chain. Rabobank Group, 47 pp. Available at: http://www.rabobank.com/content/images/Rabobank_IMW_WB_report-FINAL-A4-total_tcm43-127734.pdf.

Smil, V. 2000. *Feeding the World: A Challenge for the Twenty-First Century.* Cambridge, MA: The MIT Press, 360 pp.

Smith, A., Watkiss, P., Tweddle, G. *et al.* 2005. The validity of food miles as an indicator of sustainable development. Final report. London: DEFRA, 103 pp. Available at: http://www.igd.com/download.asp?id=3&dtid=2&did=1633.

Smith, B. G. 2008. Developing sustainable food supply chains. *Philosophical Transactions of the Royal Society B: Biological Sciences*, 363 (1492): 849–861. Available at: http://rstb.royalsocietypublishing.org/content/363/1492/849.long.

Sonnino, R. 2007. The power of place: embeddedness and local food systems in Italy and the UK. *Anthropology of Food* (6): 454. Available at: http://aof.revues.org/index454.html.

Sorensen, L. B., Moller, P., Flint, A., Martens, M. and Raben, A. 2003. Effect of sensory perception of foods on appetite and food intake: a review of studies on humans. *International Journal of Obesity*, 27 (10): 1152–1166. Available at: http://dx.doi.org/10.1038/sj.ijo.0802391.

Soussana, J. F. 2010. Quelles recherches pour adapter l'agriculture et les écosystèmes anthropisés au changement climatique? Adapting to climate change: agriculture and ecosystems., 20–22/10/2010. Clermont-Ferrand, France: Environmental Research Federation of Clermont-Ferrand, INRA, pp. 8–9. Available at: https://www1.clermont.inra.fr/urep/accae/documentation/Proceedings-ACCAE-octo2010.pdf.

Soyeux, A. 2010. La lutte contre le gaspillage. Quel rôle face aux défis alimentaires? [The battle against waste. How are we to set about the challenge of food wastage?] *Futuribles*, 362, 57–68.

Spore 2011. Sus au gaspillage! Gestion post-récolte. *SPORE*: no.152, April–May 2011. Available at: http://spore.cta.int/index.php?view=article&catid=9&id=1670&format=pdf&option=com_content&lang=fr.

Stamoulis, K. G., Pingali, P. and Shetty, P. 2004. Emerging challenges for food and nutrition policy in developing countries. *e-Journal of Agricultural and Development Economics*, 1 (2): 154–167. Available at: ftp://ftp.fao.org/docrep/fao/007/ae224e/ae224e00.pdf.

Steinberg, H., Gerber, P., Wasenaar, T. *et al.* 2006. *Livestock's Long Shadow. Environmental Issues and Options.* Rome: FAO, 390 pp. Available at: http://www.fao.org/docrep/010/a0701e/a0701e00.htm.

Stuart, T. 2009. *Waste, Uncovering the Global Food Scandal.* London: Penguin, 480 pp.

Subrahmanyam, K. V. 1986. Post-harvest losses in horticultural crops: an appraisal. *Agricultural Situation in India*, 41 (5): 339–343.

Supkova, M., Darmon, N., Martin, S. *et al*. 2010. Impact carbone de régimes alimentaires différenciés selon leur qualité nutritionnelle: une étude basée sur des données françaises. Paris: Etude co-financée par l'ADEME et l'INRA, 91 pp.

Swinnen, J. F. M. 2007. Global supply chains, standards and the poor: how the globalization of food systems and standards affects rural development and poverty. Wallingford, UK: CABI, xv + 322 pp.

Sylvander, B., Bellon, S. and Benoit, M. 2006. Facing the organic reality: the diversity of development models and their consequences on research policies. Joint Organic Congress: Farming and European Rural Development. Odense, Denmark, 30–31 May 2006, 4 pp. Available at: http://orgprints.org/8247/.

Tacoli, C. 2003. The links between urban and rural development. *Environment and Urbanization*, 15 (1): 3–12. Available at: http://eau.sagepub.com/content/15/1/3.full.pdf.

Tacoli, C. 2004. Rural-urban linkages and pro-poor agricultural growth: an overview. OECD DAC POVNET Agriculture and Pro-Poor Growth Task Team. Helsinki Workshop, 17–18 June 2004, 17 pp. Available at: http://www.oecd.org/dataoecd/25/8/36562896.pdf.

Tanumihardjo, S. A., Anderson, C., Kaufer-Horwitz, M. *et al*. 2007. Poverty, obesity, and malnutrition: an international perspective recognizing the paradox. *Journal of the American Dietetic Association*, 107 (11): 1966–1972. Available at: http://dx.doi.org/10.1016/j.jada.2007.08.007.

Tauriainen, J. and Young, F. W. 1976. The impact of urban–industrial development on agricultural incomes and productivity in Finland. *Land Economics*, 52 (2): 192–206. Available at: http://www.jstor.org/stable/3145296.

Teixeira, A. A., Chynoweth, D. P., Haley, P. J. *et al*. 2004. Prototype space mission SEBAC biological solid waste management system. *Proceedings of the International Conference on Environmental Systems. Colorado Springs*. Available at: http://papers.sae.org/2004-01-2466.

Temple, L. and Moustier, P. 2004. Les fonctions et contraintes de l'agriculture périurbaine de quelques villes africaines (Yaoundé, Cotonou, Dakar). *Cahiers Agriculture*, 13 (1): 15–22. Available at: http://www.john-libbey-eurotext.fr/fr/revues/agro_biotech/agr/e-docs/00/03/FE/DE/resume.md.

Terragni, L., Bostrom, M., Halkier, B. and Makela, J. 2009. Can consumers save the world? Everyday food consumption and dilemmas of sustainability. *Anthropology of Food* (S5): n.p. Available at: http://aof.revues.org/index6377.html.

Thomas, G. 2005. Les échanges agroalimentaires de 1992 à 2002. *Economie et statistique*, 390: 25–46. Available at: http://www.persee.fr/articleAsPDF/estat_0336-1454_2005_num_390_1_7163/article_estat_0336-1454_2005_num_390_1_7163.pdf?mode=light.

Thow, A. M. and Hawkes, C. 2009. The implications of trade liberalization for diet and health: a case study from Central America. *Globalization and Health*, 5 (5). Available at: http://dx.doi.org/10.1186/1744-8603-5-5.

Timmer, C. P. 2010. Reflections on food crises past. *Food Policy*, 35 (1): 1–11. Available at: http://dx.doi.org/10.1016/j.foodpol.2009.09.002.

Toledo, A. and Burlingame, B. 2006. Biodiversity and nutrition: a common path toward global food security and sustainable development. *Journal of Food Composition and Analysis*, 19 (6–7): 477–483. Available at: http://dx.doi.org/10.1016/j.jfca.2006.05.001.

Toutain, J. C. 1971. La consommation alimentaire en France de 1789 à 1964. *Economies et Sociétés, Cahiers de l'I.S.E.A*, V (11): 1909–2049. Available at: http://www.persee.fr/web/revues/home/prescript/article/ahess_0395-2649_1973_num_28_5_293416_t1_1212_0000_5.

Trichopoulou, A., Costacou, T., Bamia, C. and Trichopoulos, D. 2003. Adherence to a Mediterranean diet and survival in a Greek population. *New England Journal of Medicine*, 348 (26): 2599–2608. Available at: http://www.nejm.org/doi/pdf/10.1056/NEJMoa025039.

Trichopoulou, A., Kourisblazos, A., Wahlqvist, M. L. *et al*. 1995. Diet and overall survival in elderly people. *British Medical Journal*, 311 (7018): 1457–1460. Available at: http://pubmedcentralcanada.ca/picrender.cgi?accid=PMC2543726&blobtype=pdf.

Tschirley, D. L. and Jayne, T. S. 2010. Exploring the logic behind Southern Africa's food crises. *World Development*, 38 (1): 76–87. Available at: http://dx.doi.org/10.1016/j.worlddev.2009.09.008.

Tsolekile, L. P. 2007. Urbanization and lifestyle changes related to non-communicable diseases: an exploration of experiences of urban residents who have relocated from the rural areas to Khayelitsha, an urban township in Cape Town. Minithesis, Cape Town: University of the Western Cape, 76 pp. Available at: http://etd.uwc.ac.za/usrfiles/modules/etd/docs/etd_gen8Srv25Nme4_1534_1194334645.pdf.

Tukker, A., Huppes, G., Guinée, J. B. *et al*. 2006. Environmental Impact of Products (EIPRO) analysis of the life cycle environmental impacts related to the final consumption of the EU-25, *Technical Report Series*, EUR 22284 EN. Brussels: European Commission, Joint Research Centre, Institute for Prospective Technological Studies, 136 pp. Available at: http://hdl.handle.net/1887/11434.

Tukker, A., de Koning, A., Wolf, O., *et al*. 2011. Environmental impacts of changes to healthier diets in Europe. *Ecological Economics*, 70: 1776–1788. Available at: http://dx.doi.org/10.1016/j.ecolecon.2011.05.001.

Turnovsky, S. J., Shalit, H. and Schmitz, A. 1980. Consumer's surplus, price instability, and consumer welfare. *Econometrica*, 48 (1): 135–152. Available at: http://www.jstor.org/stable/1912022.

Tyler, P. S. 1982. Misconception of food losses. *Food and Nutrition Bulletin*, 4 (2): 21–24. Available at: http://www.unu.edu/Unupress/food/8F042e/8F042E05.htm.

Tyler, P. S. and Gilman, G. A. 1979. L'évaluation des pertes post-récolte. In S. Beaulieu Gingras and D. Dan Dicko, eds, *L'amélioration des systèmes post-récolte en Afrique de l'Ouest. Séminaire de Bamako*. Paris: ACCT, pp. 137–151.

UK Government 2011. Foresight Project on Global Food and Farming Futures. Synthesis Report C8: Changing consumption patterns. London: The Government Office for Science, 23 pp. Available at: http://www.bis.gov.uk/assets/bispartners/foresight/docs/food-and-farming/synthesis/11-628-c8-changing-consumption-patterns.pdf.

United Nations 2003. *World Population Prospects. The 2002 Revision*, 3 volumes. New York: United Nations, Department of Economic and Social Affairs, Population Division.

United Nations 2006. *World Population Prospects: The 2006 Revision. Population Ageing*. New York: United Nations Population Division, DESA. 8 pp. Available at: http://www.un.org/esa/population/publications/wpp2006/wpp2006_ageing.pdf.

United Nations 2007. *World Population Prospects. The 2006 Revision*. New York: United Nations, 793 pp. Available at: http://www.un.org/esa/population/publications/wpp2006/wpp2006.htm.

United Nations 2008. *World Population Prospects: The 2008 Revision Population Database*. Available at: http://esa.un.org/unpp/p2k0data.asp.

United Nations 2010. *World Urbanization Prospects. The 2009 Revision. Highlights*. New York: Department of Economic and Social Affairs, 56 pp. Available at: http://esa.un.org/unpd/wup/Documents/WUP2009_Highlights_Final.pdf.

United Nations Habitat 2003. *The Challenge of Slums: Global Report on Human Settlements 2003*. London, UK: Earthscan Publications Ltd and United Nations Human Settlements Programme (UN-Habitat), 310 pp. Available at: http://www.unhabitat.org/pmss/getElectronicVersion.asp?nr=1156&alt=1.

US-SCN 2009. *Sixth Report on the World Nutrition Situation. Progress on Nutrition*. Geneva: United Nations-System Standing Committee on Nutrition, 131 pp. Available at: http://www.unscn.org/files/Publications/RWNS6/report/SCN_report.pdf.

Vagneron, I. and Roquigny, S. 2010. Cartographie et analyse des études d'impact du commerce équitable. Montpellier, France: CIRAD, UMR MOISA. 78 pp. Available at: http://www.commercequitable.org/impact-au-sud.html.

van den Bergh, J. and Verbruggen, H. 1999. Spatial sustainability, trade and indicators: an evaluation of the 'ecological footprint'. *Ecological Economics*, 29 (1): 61–72. Available at: http://dx.doi.org/10.1016/s0921-8009(99)00032-4.

Van der Voet, E., Van Oers, L., Moll, S. *et al*. 2005. Policy review on decoupling: development of indicators to assess decoupling of economic development and environmental pressure in the EU-25 and AC-3 countries. Leiden, The Netherlands: Leiden University, Institute of Environmental Sciences (CML), 159 pp.

van der Vorst, J. G. A. J., Beulens, A. J. M. and van Beek, P. 2000. Modelling and simulating multi-echelon food systems. *European Journal of Operational Research*, 122 (2): 354–366. Available at: http://dx.doi.org/10.1016/s0377-2217(99)00238-6.

Vandecandelaere, E., Arfini, F., Belletti, G. and Marescotti, A. 2010. Territoires, produits et acteurs locaux, des liens de qualité: Guide pour promouvoir la qualité liée à l'origine et des indications géographiques durables. Rome: FAO & SinerGI, 198 pp. Available at: http://www.foodquality-origin.org/guide/giudefr.pdf.

Veerman, J. L., Barendregt, J. J. and Mackenbach, J. P. 2006. The European Common Agricultural Policy on fruits and vegetables: exploring potential health gain from reform. *The European Journal of Public Health*, 16 (1): 31–35. Available at: http://dx.doi.org/10.1093/eurpub/cki166.

Ver Ploeg, M., Breneman, V., Farrigan, T. *et al*. 2009. *Access to Affordable and Nutritious Food: Measuring and Understanding Food Deserts and Their Consequences. Report to Congress*. Washington DC: USDA, ERS, 150 pp. Available at: http://www.ers.usda.gov/Publications/AP/AP036/.

Vieux, F., Darmon, N., Touazi, D. and Soler, L. G. 2012. Greenhouse gas emissions of self-selected individual diets in France: changing the Q23 diet structure or consuming less? *Ecological Economics*, 75: 91–101. Available at: http://dx.doi.org/10.1016/j.ecolecon.2012.01.003.

Vieux, F., Soler, L. G., Touazi, D. and Darmon, N. 2013. High nutritional quality is not associated with low greenhouse gas emissions in self-selected diets of French adults. *American Journal of Clinical Nutrition* (in press).

von Braun, J., Ahmed, A., Asenso-Okyere, K. *et al.* 2008. *High Food Prices: The What, Who, and How of Proposed Policy Actions*. Washington DC: International Food Policy Research Institute (IFPRI), 12 pp. Available at: http://www.ifpri.org/sites/default/files/publications/foodpricespolicyaction.pdf.

von Braun, J. and Meinzen-Dick, R. 2009. 'Land grabbing' by foreign investors in developing countries. IFPRI Policy Briefs. Washington DC: IFPRI, 9 pp. Available at: http://www.ifpri.org/sites/default/files/publications/bp013all.pdf.

Waijers, P. M., Feskens, E. J. and Ocke, M. C. 2007. A critical review of predefined diet quality scores. *British Journal of Nutrition*, 97 (2): 219–231. Available at: http://ukpmc.ac.uk/abstract/MED/17298689.

Walter, J. and Wrightson, K. 1976. Dearth and the social order in Early Modern England. *Past and Present*, 71 (1): 22–42. Available at: http://dx.doi.org/10.1093/past/71.1.22.

Wang, J. and Sampson, H. A. 2011. Food allergy. *Journal of Clinical Investigation*, 121 (3): 827–35. Available at: http://www.ncbi.nlm.nih.gov/pubmed/21364287.

Wang, Z., Zhai, F., Du, S. and Popkin, B. 2008. Dynamic shifts in Chinese eating behaviors. *Pacific Journal of Clinical Nutrition*, 17 (1): 123–130. Available at: http://apjcn.nhri.org.tw/server/APJCN/Volume17/vol17.1/Finished/123-130-1054.pdf.

Waugh, F. V. 1944. Does the consumer benefit from price instability? *The Quarterly Journal of Economics*, 58 (4): 602–614. Available at: http://qje.oxfordjournals.org/content/58/4/602.abstract.

Webb, P. and Block, S. 2012. Support for agriculture during economic transformation: impacts on poverty and undernutrition. *Proceedings of the National Academy of Sciences*, 109(31): 12309–12314. Available at: http://www.pnas.org/content/109/31/12309.abstract.

Weidema, B. P. 2006. The integration of economic and social aspects in life cycle impact assessment. *International Journal of Life Cycle Assessment*, 11: 89–96. Available at: http://dx.doi.org/10.1065/lca2006.04.016.

WHO 1998. Preparation and use of food based dietary guidelines: report of a joint WHO/FAO expert consultation. WHO Technical Report Series, no. 880. Geneva: WHO, 116 pp. Available at: http://whqlibdoc.who.int/trs/WHO_TRS_880.pdf.

WHO 2003. Diet, nutrition and the prevention of chronic diseases: report of a joint WHO/FAO expert consultation. WHO Technical Report Series, no. 916. Geneva: WHO, 160 pp. Available at: http://www.who.int/hpr/NPH/docs/who_fao_expert_report.pdf.

WHO 2009. *Turning the Tide of Malnutrition: Responding to the Challenge of the 21st Century*. Geneva: WHO, 24 pp.

Willet, W. 1998. *Nutritional Epidemiology*. New York: Oxford University Press, xiv + 514 pp.

Willett, W. C., Sacks, F., Trichopoulou, A. *et al.* 1995. Mediterranean diet pyramid: a cultural model for healthy eating. *American Journal of Clinical Nutrition*, 61 (6): S1402–S1406. Available at: http://www.ajcn.org/content/61/6/1402S.abstract.

World Bank 2011. Missing food: the case of postharvest grain losses in sub-Saharan Africa. No. 60371-AFR. Washington DC: The World Bank. Economic and Sector Work, 96 pp. Available at: http://siteresources.worldbank.org/INTARD/Resources/MissingFoods10_web.pdf%0d.

World Bank 2011. World Bank development indicators. Available at: http://data.worldbank.org/indicator/SH.STA.MALN.ZS/countries?display=map.

World Future Council and Göpel, M. 2009. Celebrating the Belo Horizonte Food Security Programme. Future Policy Award 2009: Solutions for the Food Crisis. Hamburg: World Future Council, 20 pp. Available at: http://www.futurepolicy.org/fileadmin/user_upload/PDF/Future_Policy_Award_brochure.pdf.

WRAP 2009. Household food and drink waste in the UK. Final report. Waste & Resources Action Progamme (WRAP), Banbury, UK, 95 pp. Available at: http://www.wrap.org.uk/downloads/Household_food_and_drink_waste_in_the_UK_-_report.e4540013.8048.pdf%0d.

WRAP and WWF 2011. The water and carbon footprint of household food and drink waste in the UK. Final report, 52 pp. Available at: http://www.waterfootprint.org/Reports/Water-and-carbon-footprint-food-and-drink-waste-UK-2011.pdf.

Wright, B. 2009. International grain reserves and other instruments to address volatility in grain markets. Policy Research Working Paper, WPS no. 5028. Washington DC: The World Bank, 52 pp. Available at: http://www-wds.worldbank.org/servlet/WDSContentServer/WDSP/IB/2009/08/25/000158349_20090825154655/Rendered/PDF/WPS5028.pdf.

Wright, B. D. 1979. The effects of ideal production stabilization: a welfare analysis under rational behavior. *The Journal of Political Economy*, 87 (5): 1011–1033. Available at: http://www.jstor.org/stable/1833080.

Zamagni, A., Buttol, P., Buonamici, R. *et al.* 2009. D20 Blue Paper on Life Cycle Sustainability Analysis: CALCAS (Co-ordination Action for innovation in Life-Cycle Analysis for Sustainability). Project no.037075, 88 pp. Available at: http://www.leidenuniv.nl/cml/ssp/publications/calcas_report_d20.pdf.

Zetlaoui, M., Feinberg, M., Verger, P. and Clémençon, S. 2011. Extraction of Food Consumption Systems by Nonnegative Matrix Factorization (NMF) for the Assessment of Food Choices. *Biometrics*, 67 (4): 1647–1658. Available at: http://www.leidenuniv.nl/cml/ssp/publications/calcas_report_d20.pdf.

Abbreviations

Organisations/affiliations of experts

ADEME	French Environment and Energy Management Agency (*Agence de l'Environnement et de la Maîtrise de l'Energie*)
AFD	French Development Agency (*Agence Française de Développement*)
Agreenium	Consortium of French research and higher education bodies to build capacities on food security and sustainable development at an international level
ANR	French National Research Agency (*Agence Nationale de la Recherche*)
ART Zurich	Agroscope Reckenholz-Tänikon (Switzerland)
CAS	Centre for Strategic Analysis (*Centre d'Analyse Stratégique*)
CEMAGREF	Institute for Research in Environmental Sciences and Technologies (*Institut de Recherche en Sciences et Technologies pour l'Environnement*)
CGAAER	General Council for Agriculture, Food and Rural Areas (*Conseil Général de l'Agriculture, de l'Alimentation et des Espaces Ruraux*)
CIHEAM	International Centre for Advanced Mediterranean Agronomic Studies (*Centre International de Hautes Etudes Agronomiques Méditerranéennes*)
CIQUAL	French Data Centre on Food Quality (*Centre d'Information sur la Qualité des Aliments*)
CIRAD	Agricultural Research Centre for Development (*Centre de Coopération Internationale en Recherche Agronomique pour le Développement*)
CLCV	French Consumers Association (*Association Nationale de Consommateurs et Usagers*)
CREDOC	Research Centre for the Study and Monitoring of Living Conditions (*Centre de recherche pour l'étude et l'observation des conditions de vie*)
CSO-CNRS	Centre for Organisational Sociology–National Centre for Scientific Research (*Centre de sociologie des organisations–Centre national de la recherche scientifique*)
ENESAD	French Institute for Advanced Agronomy, Dijon (*Établissement national d'enseignement supérieur agronomique de Dijon*)

FCD	French Retail Federation (*Fédération des Entreprises de Commerce et de la Distribution*)
FNE	France Nature Environment (*France Nature Environnement*)
FNSEA	French Farmers' Union (*Fédération Nationale des Syndicats d'Exploitants Agricoles*)
GRET	Research and Technological Exchanges Group (*Groupe de recherche de recherche et d'échanges technologiques*)
IAURIF	Development and Urban Planning Institute, Ile-de-France Region (*Institut d'Aménagement et d'Urbanisme de la Région Ile-de-France*)
IDDRI	Institute for Sustainable Development and International Relations (*Institut du Développement Durable et des Relations Internationales*)
IFIP	French Pork and Pig Institute (*Institut du porc*)
IFREMER	French Research Institute for Exploration of the Sea (*Institut Français de Recherche pour l'Exploitation de la Mer*)
INRA	French National Institute for Agricultural Research (*Institut National de la Recherche Agronomique*)
INSERM	French National Institute of Health and Medical Research (*Institut National de la Santé et de la Recherche Médicale*)
IRD	Institute for Research and Development (*Institut de Recherche pour le Développement*)
ITAB	Technical Institute for Organic Farming (*Institut Technique de l'Agriculture Biologique*)
MAAPRAT	French Ministry for Agriculture, Food, Fisheries, Rurality and Territorial Planning (*Ministère de l'Agriculture, de l'Alimentation, de la Pêche, de la Ruralité et de l'Aménagement du Territoire*)
MAEE	French Ministry for Foreign and European Affairs (*Ministère des Affaires Etrangères et Européennes*)
MEDDTL	French Ministry for Ecology, Sustainable Development, Transport and Housing (*Ministère de l'Ecologie, du Développement durable, des Transports et du Logement*)
ONIRIS	National Veterinary and Agrifood School, Nantes-Atlantique (*Ecole Nationale Vétérinaire, Agroalimentaire et de l'Alimentation Nantes-Atlantique*)
SYNDIGEL	European Federation for the Commerce and Distribution of Frozen Foods and Ice Cream (*Fédération Européenne du Commerce et de la Distribution des Produits sous Température dirigée, glaces, surgelés et réfrigérés*)
WWF France	World Wildlife Fund, France

Acronyms

AFSSA	French Food Safety Agency (now ANSES) (*Agence française de sécurité sanitaire des aliments*)

ANSES	French Agency for Food, Environmental and Occupational Health & Safety
ASEAN	Association of Southeast Asian Nations (ANASE)
ASPO	Association for the Study of Peak Oil
BMI	Body Mass Index
CAP	Common Agricultural Policy
DEFRA	Department for the Environment, Food and Rural Affairs (UK)
EEA	European Environment Agency
EU	European Union
EUREP	Euro Retail Group
FAO	Food and Agriculture Organization of the United Nations
FSC label	Forest Stewardship Council Label
GATT	General Agreement on Tariffs and Trade
GEMRCN	Collective Catering and Nutrition Market Study
GHG	Greenhouse Gases
GI	Geographical Indication
IEA	International Energy Agency
IFOAM	International Federation of Organic Agriculture Movements
IFP	Intermediate Food Products
IFP	Institut Français du Pétrole
IPCC	Intergovernmental Panel on Climate Change
IRDES	French Institute for Research and Information in Health Economics
LCA	Life Cycle Assessment
MAR	Mean Adequacy Ratio
MDER	Minimum Dietary Energy Requirements
MER	Mean Excess Ratio
MSC label	Marine Stewardship Council label
NGO	Non-Governmental Organisation
NRI	Natural Resources Institute
OECD	Organisation for Economic Co-operation and Development
ONPES	French National Observatory on Poverty and Social Exclusion
PAP	Processed Animal Proteins
PNA	French National Food Programme
PNNS	French National Nutrition and Health Programme (*Programme National Nutrition Santé*)
RESEDA	Network of Professional Organisations for the Safety and Quality of Animal Foods (*Réseau des organisations professionnelles et interprofessionnelles pour la Sécurité et la qualité des Denrées Animales*)
RNI	Recommended Nutritional Intake
SIWI	Stockholm International Water Institute
SPC	Socio-Professional Category (*CSP*)
TRIPS	Trade Related Aspects of Intellectual Property Rights (*ADPIC*)

UNO	United Nations Organisation (*ONU*)
USDA	United States Department of Agriculture
USGS	United States Geological Survey
WFP	World Food Programme (*PAM*)
WHO	World Health Organization (*OMS*)
WTO	World Trade Organisation (*OMC*)

Units

eq	equivalent
g	gramme
cal	calorie
G	giga, billion
gkcal	giga kilocalories or billion kilocalories (kcal)
ha	hectare
k	kilo, thousand
kcal	kilocalories
kg	kilogram
km	kilometre
M	mega, million
BTKM	billion tonne-kilometres
ppm	part(s) per million
TOE	tonne of oil equivalent (Teq)

duALIne experts

The following experts participated in the duALIne project between November 2009 and June 2011:

Joël Abecassis, INRA
Pascale Adeline, Nestlé France
Joël Aubin, INRA, workshop leader
Christine Aubry, INRA
Dominique Auverlot, CAS, Member of Monitoring Committee
Sylvie Avallone, Montpellier Supagro
Delphine Babin-Pelliard, MAEE, Member of Monitoring Committee
Claudine Basset-Mens, CIRAD
Jérôme Bédier, FCD, Member of Monitoring Committee
Jacques Berger, IRD
Fabien Berges, INRA
Estelle Bienabe, CIRAD
Céline Bignebat, INRA
Jean-Christophe Bligny, Danone
Jean-Manuel Bluet, Nestlé France, Member of Monitoring Committee
Gérard Boivin, BEL, Member of Monitoring Committee
Jean-Pierre Boutonnet, INRA
Nicolas Bricas, CIRAD, CIRAD leader of duALIne
Christian Brodhag, Mines Saint-Etienne
Cécile Broutin, GRET
Florence Buchholzer, Directorate General for Agriculture, Member of
    Monitoring Committee
Jean-Christophe Bureau, INRA
France Caillavet, INRA
Danièle Capt, AgroSup Dijon
Patrick Caron, CIRAD, Member of Monitoring Committee
Marc Castello, Kraft

Katia Castetbon, Université Paris XIII
Armelle Champenois, INRA
Jean Chavel, Charal, Member of Monitoring Committee
Philippe Chemineau, INRA, workshop leader
Bernard Chevalier, ORGECO, Member of Monitoring Committee
Bernard Chevassus-au-Louis, Member of Monitoring Committee
Yuna Chiffoleau, INRA
Myriam Cohen-Welgryn, Danone, Member of Monitoring Committee
Bernard Collin, Charal Viandes
Paul Colonna, INRA, workshop leader
Pierre Combris, INRA, workshop leader
Georges Corrieu, INRA
Laurent Cousin, Sodexo
Joseph Culioli, INRA
Nicole Darmon, INRA
Benoît Daviron, CIRAD
Gilbert Delahaye, Casino, Member of Monitoring Committee
Francis Delpeuch, IRD Morocco
Christian Deverre, INRA
Anne-Célia Disdier, INRA
Catherine Donnars, INRA, deputy workshop leader
Bruno Dorin, CIRAD
Françoise Dosba, Montpellier Supagro
Sophie Dubuisson-Quellier, CSO-CNRS
Gérard Duc, INRA
Marie-Christine Duchamp, CIRAD
Florence Egal, FAO, Member of Monitoring Committee
Valéry Elisseeff, Society of French Farmers, Member of Monitoring
    Committee
Jean-Louis Escudier, INRA
Catherine Esnouf, INRA, INRA leader of duALIne
Patrick Etiévant, INRA
Elisabeth Faguer, IAURIF
Erwan Fangeat, ADEME
Max Feinberg, AgroParisTech
Thierry Fellmann, APCA, Member of Monitoring Committee
Léopold Fézeu, INSERM-INRA
Muriel Figuié, CIRAD
Claude Fischler, CNRS
Bernard Fostier, IFIP
Vincent Fouchier, IAURIF
Stéphane Fournier, Montpellier Supagro, deputy workshop leader

Lise Frappier, INRA
Benoît Gabrielle, AgroParisTech
Carl Gaigné, INRA, workshop leader
Gérard Gaillard, ART Zurich
Frédérick Garcia, INRA
Laure Garnier, Kraft
Frédérique Gaulard, BEL
Lucie Gauthier-Deltour, ADEME
Sophie Girard, IFREMER
Alexandre Gohin, INRA
Hervé Gomichon, Carrefour, Member of Monitoring Committee
Christophe Gouel, INRA
Michel Griffon, ANR, Member of Monitoring Committee
Jean-Yves Grosclaude, AFD, Member of Monitoring Committee
Marion Guillou, INRA, Chairman of Monitoring Committee
Hervé Guyomard, INRA, Member of Monitoring Committee
Pascale Hébel, CREDOC
Emmanuelle Henry, Ecobilan
Serge Hercberg, INSERM-INRA
Mohamed Hilal, INRA
Jean Hirschler, Normandy Chamber of Agriculture
Michelle Holdsworth, IRD
Vincent Hovelaque, ONORIS Nantes
Claire Hubert, MEDDTL, Member of Monitoring Committee
Marie-Jeanne Husset, 60 Millions de Consommateurs, Member of
    Monitoring Committee
Jean-Marc Jancovici, Member of Monitoring Committee
Pierre Janin, IRD
Sébastien Jean, INRA, workshop leader
Benoît Jeannequin, INRA
Claudine Joly, FNE, Member of Monitoring Committee
Emmanuelle Kesse, INRA
Claire Khoury, CIRAD
Eric Labouze, Biointelligence Service
Céline Laisney, MAAPRAT
Christiane Lambert, FNSEA, Member of Monitoring Committee
Jean-Louis Lambert, former professor at Enitiaa
Frédéric Lançon, CIRAD
Sylvaine Lemeilleur, CIRAD
Tristan Le Cotty, CIRAD
Sophie Le Perchec, INRA
François Letissier, Bonduelle

Danielle Lo Stimolo, SYNDIGEL
Denis Loeillet, CIRAD
Catherine Macombe, CEMAGREF
Marie-Laurence Madignier, CGAAER, Member of Monitoring Committee
Thuriane Mahé, MAAPRAT
Bernard Maire, IRD
Élodie Maitre d'Hôtel, CIRAD
Gérard Matheron, CIRAD, Chairman of Monitoring Committee
Vanessa Méry, INRA
Rémi Mongruel, IFREMER
Didier Moreau, Danone
Jérôme Mousset, ADEME
Paule Moustier, CIRAD
Eric Mugnier, Ernst and Young
Didier Nedelec, In vivo, Member of Monitoring Committee
Bénédicte Oberti, CIHEAM
Anne Pacquet, Crédit Agricole, Member of Monitoring Committee
Martine Padilla, CIHEAM
Sandrine Paillard, ANR, deputy workshop leader
Damien Paineau, Danone
Dominique Pallet, CIRAD
Laurent Parrot, CIRAD
Charles Pernin, CLCV, Member of Monitoring Committee
Alfredo Pimper, Kraft
Cécile Piot, Charal Viandes
Vincent Piveteau, INRA
Jean-Pierre Poulain, Université Toulouse II
Jean-Luc Pujol, INRA
Jean Razungles, INRA
Barbara Redlingshöfer, INRA
Catherine Renard, INRA
Michel Renard, INRA
Jean-Pierre Rennaud, Danone
Vincent Réquillart, INRA
Patrice Robichon, Pernod-Ricard
Camille Rojot, Ecobilan
Tévécia Ronzon, INRA
Marie Russel, INRA, coordinator of duALIne
Denis Sautier, CIRAD
Bertrand Schmitt, INRA
Didier Simon, AFD
Lucie Sirieix, Montpellier Supagro

Louis-Georges Soler, INRA, workshop leader
Annie Soyeux, formerly MAAPRAT, workshop leader
Christophe Soulard, INRA
Jean-François Soussana, INRA, Member of Monitoring Committee
Eric Spinnler, AgroParisTech
Philippe Steinmetz, AFD
Marketa Supkowa, UrbanFoodLab, Sustainable Food for Cities
Bruno Taupier, ITAB, Member of Monitoring Committee
Ludovic Temple, CIRAD
Daniel Tomé, AgroParisTech
Jean-Marc Touzard, INRA
Sébastien Treyer, IDDRI, Member of Monitoring Committee
Gilles Trystram, AgroParisTech
Philippe Van de Maele, ADEME, Member of Monitoring Committee
Marc Vandeputte, INRA
Roland Vaxelaire, Management Responsibility, Member of Monitoring
    Committee
Ambroise Veillon, Danone
Georges Vermeersch, Proléa
Bruno Vindel, AFD
Eric Zunino, MAAPRAT, Member of Monitoring Committee

Index

Locators in **bold** refer to figures/tables/boxed material.